Alaka Basu

The Changing Population of China

The Changing Population of China

Edited by Peng Xizhe
with Zhigang Guo

2 4 6 8 10 9 7 5 3 1

Blackwell Publishers Ltd
108 Cowley Road
Oxford OX4 1JF
UK

Blackwell Publishers Inc.
350 Main Street
Malden, Massachusetts 02148
USA

British Library Cataloguing in Publication Data
A CIP catalogue record for this book is available from the British Library.

Library of Congress Cataloging-in-Publication Data
is available for this book
ISBN 0 631 20191 2 (hbk); 0 631 20192 0 (pbk)

Typeset in Times on 10/11½ pt
By Best-set Typesetter Ltd., Hong Kong
Printed in Great Britain by Biddles Ltd, *www.biddles.co.uk*
This book is printed on acid-free paper

Contents

List of Figures

List of Tables

List of Contributors

DAI, Xingyi, Professor, Department of Environment Sciences, Fudan University, China. PhD in Economics, Fudan University, China, 1994.

GUO, Zhigang, Professor and Director, Institute of Population of Research, Renmin (People's) University of China; Professor, Institute of Sociology and Anthropology, Peking University. PhD in Population Studies, Renmin (People's) University of China, 1990.

DU, Peng, Associate Professor, Institute of Population Research, Renmin (People's) University of China. PhD in Population Studies, Renmin (People's) University of China, 1992.

HAO, Hongsheng, Professor and Deputy Director, Institute of Population Research, Renmin (People's) University of China. PhD in Population Studies, Renmin (People's) University of China, 1990.

LI, Yongping, Professor, Institute of Population Research, Peking University, China. PhD in Population Studies, University of California at Berkeley, USA, 1991.

LUI, Ping-keung, Principal Lecturer, Department of Applied Social Studies, The Hong Kong Polytechnic University. MSc in Mathematics, University of London, UK, 1974.

PENG, Xizhe, Professor and Director, Institute of Population Research, Fudan University, China. PhD in Population Studies, London School of Economics and Political Sciences, University of London, UK, 1988.

SUN, Changmin, Director and Senior Research Fellow, Institute of Population and Development Studies, Shanghai Academy of Social Sciences, China. PhD in Sociology, University of Zurich, Switzerland, 1992.

TAN, Lin, Professor and Deputy Director, Population and Development Research Institute, Nankai University, China. PhD in System Engineering, Xian Jiaotong University, China, 1990.

TANG, Shenglan, Lecturer at the International Health Division of the Liverpool School of Tropical Medicine, UK, and an Adjunct Associate Professor of the School of Public Health of Shanghai Medical University, China. PhD in Development Studies, the Institute of Development Studies at the University of Sussex, UK, 1999.

TU, Ping, Professor and Chair, Department of Marketing, Guanghua School of Management, Peking University, China. PhD in Demography, University of California at Berkeley, USA, 1989.

WANG, Guixin, Professor, Institute of Population Research, East China Normal University, China. PhD in Science (Geography), East China Normal University, China, 1996.

Wu, Zhuochun, Associate Professor at the Department of Health Statistics and Social Medicine of Shanghai Medical University, China. PhD Candidate, University of Helsinki, Finland.

XIE, Zhenming, Senior Researcher and Deputy Director, China Population Information and Research Center (CPIRC). MA in Demography, University of California at Berkeley, USA, 1985.

YE, Wenzhen, Professor, Economics Department of Xiamen University, China. PhD in Sociology, Utah University, USA, 1991.

ZENG, Yi, Senior Research Scientist, Center for Demographic Studies and Departmemt of Sociology, Duke University, USA; Professor, Institute of Population Research, Peking University, China; Distinguished Research Scholar, Max Planck Institute of Demographic Research, Germany. PhD in Demography, Brussels Free University, Belgium, 1986.

ZHAI, Zhenwu, Professor and Deputy Director, Institute of Population Research, Renmin (People's) University of China. PhD in Population Studies, Renmin (People's) University of China, 1990.

ZHONG, Fenggan, Professor, Institute of Population Research, Zhongshan University, China. MA in Economic Geography, Zhongshan University, China, 1981.

ZUO, Xuejin, Vice President and Senior Research Fellow, Shanghai Academy of Social Sciences, China. PhD in Economics, University of Pittsburgh, USA, 1989.

1   Introduction

Peng Xizhe

Located in East Asia, China is the world's third largest country by area and its largest by population. In 1998 China had an estimated population of 1.25 billion; the population density was about 130 persons per square kilometre, but with a very uneven geographic distribution. The majority of the Chinese population live in the eastern part of the country, while the inland mountainous regions are sparsely populated. China is a multi-ethnic country, with 56 recognized ethnic nationalities. Han people account for about 92 per cent of the total Chinese population, while the other 55 nationalities share the remaining 8 per cent. Most of these minority groups are distinguished from the Han population by language or religion rather than by racial characteristics.

China has a long and rich cultural tradition; it gave birth to one of the world's earliest and most continuous civilizations and has a recorded history that dates from some 4,000 years ago. Confucian ideology dominated Chinese society for centuries. By the nineteenth century, however, China had become a politically and economically weak nation. This situation was attributable not only to invasion and domination by foreign powers, but also to the fact that the Chinese people were, as Dr Sun Yat-sen, the leader of the republican revolutionary movement, deplored, 'unbound as loose as sand'. After long years of struggle for independence and self-respect, the People's Republic of China was founded in 1949. This stands as one of the most important events in Chinese history (Butler, 1993). In a remarkably short period of time, fundamental social and economic changes transformed China into a new nation. However, striving for modernization was not without its difficulties. There were ups and downs. Among the most influential events during this process were the Great Leap Forward and its attendant crisis 1958–62, and the Great Cultural Revolution that plunged the country into turmoil during the period 1966–76.[1]

It was the third plenary meeting of the 11th Central Committee of the Communist Party of China (CPC), convened in December 1978, which marked the beginning of the emancipation of thought in modern China and started the whole process of economic reform. The spirit of China's reform was symbolized by the saying of the late Deng Xiaoping: 'It does not matter if a cat is black or white as long as it catches the mouse'. A cover story in the *Beijing Review* declared that the Reform and Opening programme and the modernization drive have involved both a continuous emancipation of the mind and constant exploration and new practices (Lin and Dai, 1998). In 1992, after more than a decade of pilot experiments, the market finally came to be regarded as the primary force directing the allocation of resources. The 'invisible hand' of the market gradually replaced the 'iron hand' of a centrally planned economy. Today, Chinese people are fully aware that the fundamental tasks at present are to develop China's economy and improve its people's living standards.

Following the guidelines laid down by Deng Xiaoping and the third plenary meeting of the 11th CPC Central Committee, some fundamental reforms have been carried out. A new rural household responsibility system was first introduced in 1978 in Anhui province and then spread widely across the country. In 1983 the people's commune system, which was set up in 1958 and characterized by being 'large in size and collective in nature', was formally abandoned. At the same time the system of unified and fixed state purchase of agricultural products was abolished. These measures of rural reform have drastically increased agricultural production and helped to solve the food problems of China's 1.2 billion population. Rural reform has also freed millions of rural labourers from agricultural work and had a great impact on the development of the rural economy. Rural enterprises, known as Township and Village Enterprises (TVEs), have grown at a spectacularly fast rate, in terms of both numbers and output values. These TVEs presently employ more than 100 million people, or about one quarter of the total rural labour force.

One of the major strategic approaches of China's reform is, in the words of Deng Xiaoping, to 'make some people rich first, so as to lead all people to wealth'. Starting from the early 1980s, four special economic zones, namely Shenzhen, Zhuhai, Shantou and Xiamen, were set up in succession. This was followed by the opening of 14 coastal port cities, the development of Shanghai's Pudong New Area, the establishment of Hainan Province, and the opening of the Yangtze River Delta and the Pearl River Delta, etc. Successfully attracting large-scale foreign investment and benefiting from privileged treatment under government policy, these areas have experienced double-digit economic growth during the last two decades. Nowadays China is much more open to the outside world, and globalization is really on the way.

Inland areas, on the other hand, have lagged behind. In other words, the gap between the rich and the poor has enlarged both as between regions and within particular localities (see table 1.1). To maintain social stability and safeguard the reform process, many efforts have been made to reduce the gap and to ensure that people benefit equally from development. Actions to alleviate poverty have been undertaken since the mid-1980s by both the Chinese government at all levels and international organizations. The number of the rural population classified as poor decreased from 300 million in the early 1980s to 48 million in 1998. Although the poverty line criterion in China's countryside is quite low, starvation and famine no longer occur. The urban basic-living-guarantee system has also been established to help those urban residents who live below the poverty line solve the crucial problems of subsistence. So far, all the Chinese cities and about 81 per cent of the counties have established the guarantee system, and millions of urban residents have received relief from it. However, social diversification and segregation still need to be given more attention.

Huge investment has been poured into the transportation sectors. The total length of railway line had reached 57,600 kilometres by the end of 1997, 18.5 per cent more than the 1978 figure. Highways extended to 1.226 million kilometres, up 37.7 per cent; air routes soared to 1.425 million kilometres, a 9.7 fold increase during the same period. In 1997 about 8 per cent of the population had their own telephones, with the figure rising to 26 per cent for urban residents. Optical fibre lines have been laid to all provincial cities. The handling capacity of long-distance telephone switches grew from 11,000 lines in 1978 to 4.368 million in 1997. The effects of the relaxation of the state monopoly in the transportation and communication sectors are particularly apparent.

The urbanization process has accelerated since the beginning of the reform. The number of cities increased from 193 in 1979 to 668 in 1997, with most of the new cities growing out of the rapidly expending counties along the coastal regions. Cities have become the backbone of the national economy. Along with this rapid urbanization, rural–urban migration on an unprecedentedly large scale has become a social phenomenon with profound socio-economic consequences. The so-called 'floating population' produced by it amounts to 80 million and can be found almost everywhere in China.

The average income of Chinese people in 1997 was 14 times that of 1979, a 212 per cent increase in real terms, according to the State Statistics Bureau (SSB) of China.[2] Between 1979 and 1984, the income of farmers grew more rapidly that that of urban residents due mainly to massive rural reform. However, conditions have become more favourable for urban residents since 1985 as urban reform has gained momentum. The average annual urban income was 2.7 times that of farmers, even though the farmers' average annual income soared from 134 yuan in 1978 to 2,090 yuan in 1997, an annual increase of 8 per cent on average.

Similar changes have taken place in almost all aspects of life. By the end of 1997, for every 100 Chinese families, there were 100.48 colour televisions, 72.98 refrigerators, 165.74 electric fans and 89.12 washing machines. Meanwhile living space per capita in urban areas soared from 3.6 square metres in 1978 to 8.8 square metres in 1997. Investments in housing construction amounted to 3,241.7 billion yuan, accounting for 23 per cent of total fixed-assets investment over the last two decades. The improvement for farmers was even greater: per capita living space rose from 8.1 square metres to 22.46 square metres during the last two decades, an increase of 280 per cent.

For a long time, jobs were assigned by the state to individuals, with neither workers nor the enterprises having any say in the matter. The 'iron rice bowl' employment system ended in the early 1980s when contractual employment was initialized nationwide. The transformation of the centrally planned economy towards a market-orientated one has made the labour force increasingly mobile, and people have been given much more freedom to choose their job since then. By the end of 1997 there were 68 million people working in the private sector, whereas there were only 150,000 in 1987. Among these 68 million, 4.7 million were working in shareholding enterprises, and another 5.8 million in foreign-funded firms, and 13.39 in private enterprises (*Beijing Review*, 1998). From 1989 to 1997 the number of private enterprises grew from 90,581 to 960,726, with an annual growth of 34.3 per cent.

Over the same period state-owned enterprises (SOEs) became an increasingly serious issue for China's reformers. The poor economic performance of the SOEs put great pressure on the establishment of a labour market and the reform of the social security system. Unemployment has become a serious social and economic problem affecting millions of members of the urban labour force and their families. It is estimated that there were roughly 15 million unemployed people in urban areas at the end of 1998, accounting for 8 per cent of the total urban work force.[3] Various measures have been taken to carry out reform of the SOEs and to build new institutional mechanisms to help redundant workers. Between 1983 and 1996 more than 70 million workers found new jobs through measures of this kind.

After years of discussion and experiments a comprehensive social security system, including pensions, unemployment benefit and medical insurance, has gradually emerged. This system is facing great difficulties in its early stages. Since SOEs and other collectively owned enterprises remain the main financial resource of the system, there is little room for further increase in the payment rate that is already as high as 35–45 per cent of the overall payroll of these enterprises. On the other hand, demands for fund allocation have exceeded the capacity of the system and will continue to grow. Among many factors, population ageing, rising uncertainty in employment and soaring medical costs make the restructuring of the social security system one of the most difficult and sensitive of the reform measures, since it affects the life of every Chinese person. So far the system has mainly worked in urban areas, and the coverage of the rural population is very limited. There is a long way to go to setting up a social security system for every Chinese citizen.

Reform in other social sectors has accelerated. Education is another area that has benefited from the reform process. A nationwide campaign to enforce nine-year compulsory education has been launched. The whole society has been mobilized, through volunteer activities such as the 'Project Hope',[4] to assist school dropouts from poor families to complete their elementary education and to improve school facilities in impoverished rural areas. Reform of higher education, including the tuition-fee system, is expected to boost investment in human resources and make a contribution to China's sustainable development. In fact, China had more than 6.3 million college students in 1997 while the figure was only 1.2 million 20 years ago. Young people are the group that has benefited most from the reform and opening up. Compared to the older generations, young people nowadays have more opportunities for education, and have more freedom to choose their jobs.

We do not intend, and are also unable on grounds of space, to list all the social changes that have occurred in the last two decades. We simply wish to provide some background information for our readers to enable them to have a better understanding of the main theme of this book, the changing population of China. As the most populous country in the world, China's struggle for population control has attracted wide attention from international society both for its remarkable achievements and for its approach and system of measurement as well. There are shortcomings in any mass programme, and in some cases these shortcomings have had serious negative impacts. Nevertheless, China's population programme has brought about dramatic changes in people's fertility behaviour within a relatively short period and successfully slowed down the country's rapid population growth. It has been estimated that China might already have had a population of 1.5 billion if the family planning programme had not been carried out.

China's achievement in population control has also played an important role in slowing down the growth of the world's population as a whole, owing mainly to the large size of China's population. The average growth rates for Asia excluding China were 2.2 and 2.1 per cent for the periods 1970–80 and 1980–90, respectively –0.2 and 0.3 percentage points higher than the equivalent rates including China.[5] China's contribution to the stabilization of the world population should be fully acknowledged.

Figure 1.1 presents the time trend of China's demographic transition. The marked ups and downs during the period 1958–63 are conspicuously abnormal and resulted from the Great Leap Forward, as mentioned earlier. Except for that period, China witnessed a sharp and continuous mortality decline in the early 1950s, and the crude death rate has been maintained at a very low level of around 7 per 1,000 since the middle 1960s. The fertility decline started much later, in early 1970s, but proceeded at a strikingly fast rate. The crude birth rate declined from over 33 per 1,000 to less than 18 per 1,000 during this decade, while the total fertility rate dropped from 5.8

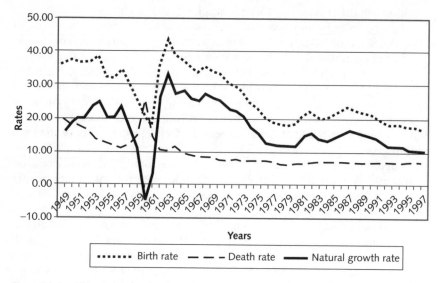

Figure 1.1   China's birth, death and natural growth rates, 1949–1997

in 1970 to 2.8 in 1979. Fertility level fluctuated during the 1980s, and resumed its general trend of decline in the 1990s. So far, the total fertility rate in China is estimated to be around 2 and is expected to remain at this level for the coming decade.

China's experience indicates that demographic transition can occur in a society with a relatively low level of socio-economic development but with strong, government-supported programmes for public health and family planning, etc. This confirms what Teitelbaum stated in 1975, that there is no evidence to show that any threshold level of development is necessary for demographic transition to happen (Teitelbaum, 1975). In addition, the international transfer of modern technology and operational systems for public health and birth control programmes has helped China's demographic transition to progress with unprecedented speed. However, it is also true that China's demographic transition benefited greatly from China's general development. In other words, if government support initiated the decline in mortality and fertility, it was socio-economic development that made such changes sustainable.

As we have already discussed, economic reform in China has brought about great improvement in the Chinese people's quality of life, and note-worthy changes in the freedom and opportunities available to individuals, thus creating a favourable social environment for China's population programme. This general development could be seen clearly from the time trend of China's Human Development Index (HDI) that was prepared by the United Nations Development Programme (UNDP). The index is based on three indicators: longevity, as measured by life expectancy at birth; educational attainment, as measured by a combination of adult literacy and the combined gross primary, secondary and tertiary enrolment ratios; and standard of living, as measured by real GDP per capita (PPP$).[6] According to UNDP estimates, China's Human Development Index was 0.51 in 1975. It has since then increased steadily and continuously up to 0.554, 0.588, 0.624 and 0.701 in 1980, 1985, 1990 and 1997, respectively. At present, China has achieved medium-level human development, ranking 98th among all the 174 countries or regions in the world for which the UNDP calculated the HDI.

However, the development process is not progressing evenly across the country. There are regional variations. Table 1.1 shows the Human Development Index and its component indicators, together with the Crude Birth Rate (CBR), for each of China's provincial units. There are 13 provinces where the HDI exceeds the national index, while it is below the national average in the remaining 17 provinces. The HDI value for three major Chinese cities, Shanghai, Beijing and Tianjin, and for Guangdong province has already reached a level that the UNDP classifies as high human development. But the level in Qinghai, Xizang and some other inland provinces is very low. The extremely low value of the HDI in Xizang (Tibet) is mainly caused by its low education index.

There is a close relation between these human development indices and the birth rate. The correlation coefficient between HDI and CBR is as high as −0.813. It is common for low fertility and high life expectancy to be observed in those provinces with a higher GDP index, while a relatively high fertility rate is linked to a low level of socio-economic development.[7] There are a few special cases, such as Guangdong province, where a high level of socio-economic development is not necessarily accompanied by a low birth rate. The composition of the ethnic population, level of urbanization, local cultural tradition, efficiency of government administration, even the natural environment, all contribute to the regional variations in the demographic patterns.

It is not our intention to focus on narrowly defined population matters in China, as many books and journal articles are already available to readers. Instead, we take a multidisciplinary approach and try to cover a wide range of socio-economic issues centred on or closely linked to population dynamics. In order to reach wider readership, this book tries to be more informative and provides limited technical details. It is hoped that readers of this book will find it a comprehensive source of information about the social contexts and consequences of China's rapid population change.

The book begins with assessments of China's population distribution mortality and fertility patterns, and related population policy. These chapters are followed by others dealing with the consequences of those events, namely age and sex structures, population ageing, marriage and family

Table 1.1 Human development index (HDI) and crude birth rate (CBR) by province, China, 1995

Rank	Province	HDI	Life expectancy index	Education index	GDP index	CBR (0.000)
1	Shanghai	0.885	0.84	0.85	0.969	5.75
2	Beijing	0.876	0.81	0.86	0.960	7.92
3	Tianjin	0.859	0.80	0.83	0.954	10.23
4	Guangdong	0.814	0.80	0.79	0.850	18.10
5	Zhejiang	0.785	0.79	0.75	0.814	12.66
6	Jiangsu	0.760	0.79	0.77	0.724	12.32
7	Liaoning	0.756	0.76	0.80	0.708	12.17
8	Fujian	0.729	0.76	0.72	0.709	15.20
9	Shandong	0.704	0.77	0.74	0.604	9.82
10	Heilongjiang	0.676	0.72	0.78	0.526	13.23
11	Hainan	0.674	0.79	0.75	0.488	20.12
12	Hebei	0.670	0.78	0.77	0.464	13.93
13	Jilin	0.659	0.72	0.80	0.451	12.90
14	Shanxi	0.627	0.74	0.79	0.352	16.60
15	Xinjiang	0.619	0.67	0.75	0.438	18.90
16	Henan	0.618	0.75	0.74	0.358	14.41
17	Hubei	0.609	0.71	0.73	0.388	16.18
18	Guangxi	0.605	0.74	0.75	0.332	17.54
19	Anhui	0.600	0.75	0.72	0.328	16.07
20	Hunan	0.592	0.71	0.75	0.320	13.02
21	Sichuan	0.582	0.70	0.74	0.308	17.08
22	Inner Mongolia	0.578	0.70	0.74	0.296	17.23
23	Jiangxi	0.577	0.70	0.73	0.327	18.94
24	Ningxia	0.571	0.72	0.67	0.323	19.28
25	Shaanxi	0.570	0.72	0.73	0.259	15.93
26	Yunnan	0.526	0.65	0.64	0.289	20.75
27	Gansu	0.514	0.71	0.62	0.216	20.65
28	Qinghai	0.503	0.61	0.57	0.326	22.01
29	Guizhou	0.594	0.67	0.64	0.172	21.86
30	Xizang	0.391	0.58	0.36	0.226	24.90

Sources:
For all HDI indices, UNDP China Country Office, *China Human Development Report*, 1998.
For crude birth rate statistics, State Statistical Bureau, PRC, *China Statistical Yearbook*, 1996.

patterns. Other topics covered include education, employment, gender issues, population distribution and urbanization, internal and international migration, ethnic populations, health care, environment and population projections. The population issue of Hong Kong is discussed in a special chapter. One of the shortcomings of the book is that the demographic dynamics of Taiwan and Macao are not presented due mainly to the unavailability of data.

We must realize that Chinese society has been undertaking a fundmental change with unprecedented speed and depth. Although our authors in writing their chapters have tried to use the latest information available to them, some data may have become outdated when this book finally appears in the bookshops. For example, population information for Chongqing metropolitan area is in most cases not separated from that for Sichuan province, even though the former has been an independent province-level administrative unit since 1996. Nevertheless, we are confident that the overall information provided by this book is valid and accurate.

The authors of this book all belong to the new generation of Chinese scholars. Most of them have studied abroad, which is a result of China's open-door policy and reform. Modern demographic studies in China can be traced back to the 1920s and 1930s. However, it was the later 1970s that marked the beginning of the rapid recovery of the discipline. Many population research institutes were established and new students were recruited. The relationships between research institutes and government agencies have been exceedingly close, as the latter need academic consultations to form population policies and programmes.[8] Young scholars, including many of the authors of this book, were given the opportunity to study abroad in famous academic institutions with the support of the Chinese government and international organizations such as the United Nations Population Fund. They have now become the backbone of China's demographic community and have worked extensively on population-related issues.

Editors' Acknowledgements

As editors, we would like to take this opportunity to express our gratitude to all the people involved in the publication of this book. In particular, we are indebted to the contributors to this book for their collaboration and patience. Dr Tu Ping took part in the early preparation of the book. His important contribution is highly appreciated. Meanwhile, special thanks are given to Louise Spencely and Tessa Harvey of Blackwell Publishers. Without their initiative and continuous support, this book would never have been published. Thanks also to copy-editor Stephen Curtis and desk editor Tessa Hanford, and to Zoe Oxaal, for their editorial assistance.

Notes

1. There is a good deal of literature covering these events. For instance, MacFarquhar, Roderick, *The Origins of the Cultural Revolution*, New York: Oxford University Press. For the demographic consequences of the Great Leap Forward in China's provinces, see Peng Xizhe, 'Demographic Consequences of the Great Leap Forward in China's Provinces', *Population and Development Review*, vol. 13, no. 4, 1987, pp. 639–70.
2. Data materials here and subsequently are mainly derived from various statistical reports by the State Statistics Bureau of China, unless otherwise specified. Those interested in the facts of China's development are also referred to a 22-article series on China's achievements in reform and opening up in the last 20 years published in the *Beijing Review*, October 1998.
3. *Beijing Review*, 1–7 March, 1999.
4. A volunteer and charitable project launched in 1989 aiming to help poor rural areas improve educational opportunities for young people from poor families. Chapter 10 of this book, which deals with education, provides more information on this project.
5. US Bureau of the Census (1991), Report WP/91, *World Population Profile: 1991*. See also UN (1992), *World Population Monitoring 1991*, Population Studies, No. 126, ST/ESA/SER.A/126.
6. The Human development Index (HDI) value ranges from 0 to 1. The HDI for a country show the distance it has already travelled towards the maximum possible value of 1 and also allows comparisons with other countries. See the chapter on Human Development Indicators and the Technical Note in the *Human Development Report, 1999*, UNDP.
7. This will be further elaborated in the relevant chapters of this book.
8. Susan Greenhalgh has provided a detailed review of development demography in China. See Greenhalgh, Susan, 'Population Studies in China: Privileged Past, Anxious Future', in *The Australian Journal of Chinese Affairs*, 24 July 1990.

References

Beijing Review, Various issues, 1998 and 1999.
Butler, O. C., Salter, C. L. and Tan, K. C. (1993), China. *Microsoft Encarta*™ 1993 Microsoft Corporation.
Lin, Liangqi and Dai, Xiaohua (1998), Persistently emancipating the mind. *Beijing Review*, 15–21 June, 9–12.
Teitelbaum, Michael (1975), Relevance of demographic transition theory for developing countries. *Science*, 188, 422.
United Nations Development Programme (1998), *China: Human Development Report*.

2 The Distribution of China's Population and its Changes

Wang Guixin

The Basic Characteristics of China's Population Distribution

The population of China is the largest in the world, and is distributed extremely unevenly (see table 2.1). In 1996 the population of Sichuan province,[1] which was larger than that of any other province in China, reached about 112 million,[2] 9.40 per cent of the national total. The next largest were the populations of Henan and Shandong provinces, both of which exceeded 87 million and accounted for 7.70 and 7.32 per cent of the national total, respectively. That the total population of these three provinces accounted for 25 per cent of the national total shows their importance in China. The populations of Tibet, Qinghai and Ningxia were all very small, accounting for only 0.20, 0.39 and 0.44 per cent of the national total, respectively. Together they form about 1 per cent of the national total, slightly more than that of Beijing (0.9 per cent) and slightly less than that of Shanghai (1.1 per cent). The largest population, that of Sichuan province, was 47 times greater than the smallest, that of Tibet.

The highest population density measured in 1996 was in Shanghai: 2,138 persons per square kilometre. Shanghai was followed by Tianjin, Jiangsu and Beijing, with 792, 671 and 661 persons per square kilometre, respectively. It is worth noting that Jiangsu province had a higher population density than Beijing municipality. In some less populated provinces or regions such as Tibet, Qinghai and Xinjiang, however, the population densities were less than 10 persons per square kilometre, Tibet having a density of just 2 persons per square kilometre. There were great disparities between the population densities of the various provinces in China. The highest population density, in Shanghai, was about 1,100 times greater than the lowest, in Tibet. The population density of Jiangsu province was also 340 times that of Tibet, showing that the difference in population density between provinces was even greater than the difference in the population size.

The population distribution was also uneven as between the eastern, central and western zones of China. For example, the land areas of the eastern, central and western zones[3] account for 13.91, 29.78 and 56.30 per

Table 2.1 The distribution of China's population by provinces, 1996

Provinces	Population (10,000)	Population density (person/square km)
Eastern zone	49,230.11	377
Beijing	1,083.23	661
Tianjin	902.43	792
Hebei	6,461.03	335
Liaoning	4,056.78	266
Shanghai	1,304.43	2,138
Jiangsu	6,908.13	677
Zhejiang	4,400.09	434
Fujian	3,210.61	265
Shandong	8,747.05	568
Guangdong	6,896.77	386
Guangxi	4,545.50	192
Hainan	714.06	211
Central zone	42,924.74	154
Shanxi	3,059.21	196
Inner Mongolia	2,263.00	20
Jilin	2,579.14	134
Heilongjiang	3,605.10	79
Anhui	6,053.98	431
Jiangxi	3,981.03	237
Henan	9,203.06	554
Hubei	5,776.37	313
Hunan	6,403.85	302
Western zone	27,391.39	51
Sichuan	11,238.23	196
Guizhou	3,459.54	198
Yunnan	3,909.38	101
Tibet	239.30	2
Shaanxi	3,457.69	169
Gansu	2,427.83	63
Qinghai	462.65	6
Ningxia	521.21	100
Xinjiang	1,657.56	10

Source: *China's Population Statistical Yearbook 1997*, SSB, China Statistical Press, 1997, p. 413. The data for the area of every province was taken from the data of the *Map of China's Land Resource (1:1,000,000)*, edited by the China Scientific Academy and the Committee of Comprehensive Observation of Natural Resources of State Planning Committee, China People's University Press, 1991, p. 14. Military personnel are not included in provincial populations.

cent of the national total area, respectively, increasing from east to west. The populations of these three zones, as proportions of the national total, are 41.18, 35.91 and 22.91 per cent, respectively, thus falling from east to west. Accordingly, population density decreases from 373 persons per square kilometre in the eastern zone, to 152 persons per square kilometre in the central and to 51 persons per square kilometre in the western zone. Along a line joining Shanghai in the east of China, where the Changjiang River reaches the East China Sea, to Tibet in the west part of China on the Qinghai–Tibet Plateau, there are six provinces or autonomous regions, Shanghai, Jiangsu, Anhui, Hubei, Sichuan and Tibet. Their population densities, as measured in 1996, decrease from 2,138 to 677, 431, 313, 196 and finally 2 persons per square kilometre, respectively, showing a downward tendency in population density from east to west. In the eastern zone, however, the population distribution is also unbalanced. The population densities of the three municipalities of Beijing, Tianjin and Shanghai, and of Jiangsu and Shandong provinces are higher than those of the other eastern provinces. If China is divided into two parts by the Aihui–Tengchong Demarcation Line (Hu Huanyong, 1990), China's population is overwhelmingly in the southeast part of China (see figure 2.1).

The total population of mainland China had increased from 542,000,000 in 1949 to 1,236,260,000 by the end of 1997, an increase of 694 million, or 128.1 per cent, over a period of 48 years. The distribution of population in

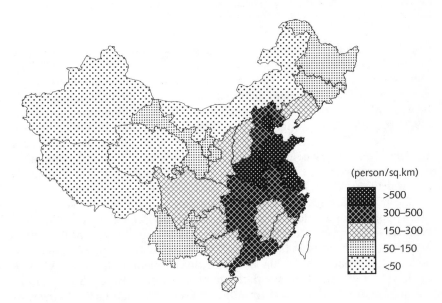

(person/sq.km)

>500
300–500
150–300
50–150
<50

Figure 2.1 Distribution of China's population, 1996

Source: As table 2.1.
Data for Taiwan. Hong Kong and Macao are not included.

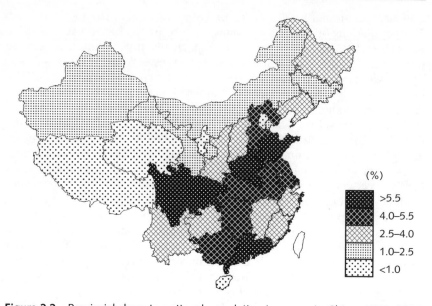

Figure 2.2 Provincial share to national population increases in China, 1950–1996

Source: China's Population Statistical Yearbook 1997 as for table 2.1. Collection of historical statistical data by province, autonomous region and municipality in China from 1949 to 1989 by SSB. China Statistical Press, 1990; Population of some provinces in 1950 was estimated by the author. Data for Taiwan, Hong Kong and Macao are not included.

China changed too. The population of each province increased substantially but the larger the population was, the more it increased (see figure 2.2). During the period 1950 to 1996, Sichuan province, whose population was the largest in China, also had the largest increase in population, 54,082,300. It was followed by Henan province (49,210,600), Shandong province (49,210,6000), Guangdong province (40,470,500) and Jiangsu province (33,251,300). The total increase in population of these five provinces accounted for 35 per cent of the increase in the total national population over the same period. Even the Tibet Autonomous Region, whose population was the smallest, had an increase of 1,370,000.

The population growth rates in boundary and remote provinces with small populations and low densities, together with Beijing, were higher than those in other provinces. If we divide China into three zones of high, medium and low latitude (see figure 2.3), the population growth rate in the high-latitude zone was the highest while that of medium-latitude zone was the lowest, and that of the low-latitude zone was in-between. So the population growth rate decreased from the remote and boundary provinces to the inner provinces over the whole country, with the exception of Sichuan province. However, Jiangsu and Shandong provinces in the eastern zone, both of which had large populations and high densities, had the lowest rates

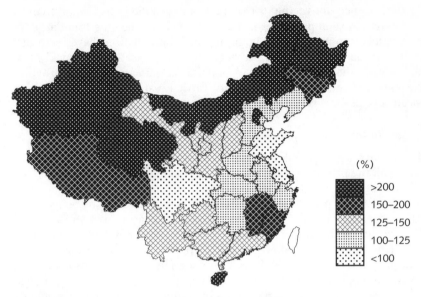

(%)
■ >200
▨ 150–200
▦ 125–150
▥ 100–125
▨ <100

Figure 2.3 Population growth rates by province in China, 1950–1996
Source: As figure 2.2.

of population increase. Among all provinces, the growth rate of Beijing's population was the largest, more than 430 per cent over a period of 47 years. Growth rates in the boundary and remote provinces such as Ningxia, Heilongjiang, Inner Mongolia, and Henan, were also high, ranging from 203 to 314 per cent. In contrast, Sichuan province, whose size and density of population were both the largest in China, had only a 92.8 per cent growth rate. Likewise, in other densely populated provinces such as Shandong and Jiangsu in the eastern zone, the growth rates were lower than the national average. For example, those of Shangdong and Jiangsu were only 88.5 and 92.8 per cent, respectively.

It should be pointed out that the population increase in the eastern area, however, has accelerated since the economic reform of the late 1970s. For example, the growth rate of the population of Guangdong province, the spearhead of economic reform in China, reached 31.93 per cent during 1980–96, just less than that of Ningxia (39.36 per cent) which was the highest in China at the same period. The growth rates of population were 78.37 per cent in the eastern zone, 90.80 per cent in the central zone and 87.90 per cent in the western zone from 1950 to 1980; while from 1980 to 1996 they were 22.06 per cent in the eastern, 22.01 per cent in the central and 21.67 per cent in the western zone. This indicates that the population has grown faster in the post-reform than in the pre-reform period in the eastern zone, so that it has become the fastest-increasing zone.

As mentioned above, the distribution of population has tended to even out slightly between provinces in the past 50 years since the foundation of the People's Republic, for the large provincial populations have increased slowly and the less populated provinces have increased more quickly. But since reform, the differing rates of population increase in every province have enlarged the provincial disparity further, causing the population to concentrate in the eastern zone.

Changes in population density

With the increase in population size mentioned above, the population densities of all the provinces have risen, while regional disparities have tended to decrease, similar to the pattern of change in population size. As shown in figure 2.4, China's population distribution was characterized by being dense in the east and sparse in the west, decreasing from east to west in 1950. And this situation still remained the same in 1996 (as shown in figure 2.1), which means few changes have taken place in the population distribution of China in the past 50 years. In the 1930s Professor Hu Huanyong established a demarcation line from Aihui in the northeast part of China to Tengchong in the southwest part of China and found the population of China was mainly concentrated to the southeast of this line (Hu Huanyong, 1990). The population distribution has undergone few changes in the past 70 years except for the rise of population density in all the provinces in China.

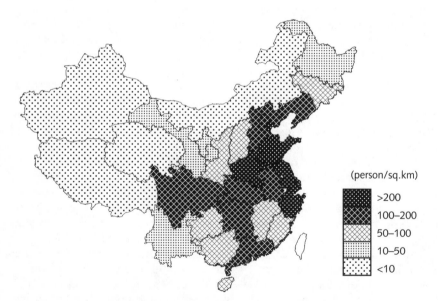

Figure 2.4 Distribution of China's population, 1950
Sources: As table 2.1 and figure 2.2.

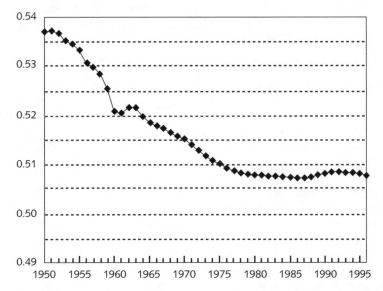

Figure 2.5 Change in the population–concentration index in China, 1950–1996
Source: As table 2.1 and figure 2.2.

Changes in the concentration index of population

Population distribution is a result of the concentration and dispersion of population, the former causing disparities in population distribution while the latter lessens them. The curve of the concentration index of population, as shown in figure 2.5, indicates that the changes in China's population distribution in the years since the founding of new China have had several characteristic features.

First, the population distribution has tended to become a bit more balanced as an overall result of natural environment and economic factors etc. during the past 50 years. The population concentration indexes decreased from 0.5371 in 1950, 0.5210 in 1960, and 0.5082 in 1990 to 0.5077 in 1996. The reduction of the population–concentration index means a trend towards an even distribution of China's population, although the general unbalanced distribution pattern remains.

Second, there have been different changes of population distribution in different periods of Chinese history. There have been two stages in the past 50 years. The first stage was from the founding of the People's Republic in 1949 to the late 1970s, during which time the population was mainly dispersing and tending towards a more even distribution. The second stage was from the late 1970s to the 1990s, when population distribution showed a reverse tendency and the population was concentrating in the east of China. Therefore, the disparity in population distribution has increased since the mid-1980s. The extent of variation, however, was relatively small and had little influence on the overall distribution of population in China.

The Main Factors Influencing Population Distribution and its Changes

China's population distribution has been deeply influenced first by physical and environmental conditions. As a result of the topography, low in east and high in west, and of the climate of the southeast monsoon, there are great differences in environmental conditions between the eastern and western zones. These have caused the great difference in population distribution between them. The eastern zone, with its good physical conditions, moist climate and arable land, can feed more people using the same land area, and accordingly has a larger population density than either the central or western zone. The Aihui–Tengchong population demarcation line closely matches the climatic demarcation line by which China's land is divided into two parts: the arid and semi-arid area and the humid area.

Second, the distribution of China's population is closely related to regional levels of economic development. The disparity in economic development between east and west, which is also affected by the physical environment, has had an important impact on the distribution of population between them. The level of economic development decreases from east to west, and population density shows the same trend. The results of a step regression for the relationship between population distribution and a number of variables relating to physical environmental conditions and level of economic development indicate that one natural variable, the percentage of the land at an elevation of less than 100 metres, and two economic variables, land productivity[4] and the proportion of industrial output value, contribute 98 per cent to the variation in China's population distribution. That is to say that China's population distribution is mainly determined by these three variables.

Further study shows that the influence of physical environmental conditions is the determining factor, contributing 80 per cent to the disparity in population distribution. The stability of the physical environmental structure, the climate and the topography has determined the stable situation in population distribution. Nevertheless, although it has had less impact on China's population distribution than natural conditions, the level of economic development, which has changed quickly and frequently as economic development has taken place, is also an important factor. For example, the economic reforms of the late 1970s brought in the strategy for advancing coastal economic development, the eastward transposition of the national economic development focus, and rapid economic development in the coastal area. Since that time China's population distribution has tended towards concentration in the east, a quite different trend from the dispersion of population before the economic reform.

An Evaluation of China's Population Distribution and its Changes

China's population distribution is undoubtedly very unbalanced, but this does not mean it is irrational. On the basis of the analysis in the previous section, we could say that population distribution in China is fairly reasonable, although uneven. With respect to natural environmental conditions, most central zone provinces have an excess population-carrying capacity and less population pressure, especially Heilongjiang and Jilin provinces because of their plentiful land resources. Those with insufficient population-carrying capacity and more population pressure are mainly in the western zone, the Xinjiang Autonomous Region being an exception because it also has plentiful land resources. With respect to economic development levels, the provinces with excess population-carrying capacity and less population pressure are mainly in the eastern zone, especially Shanghai. And provinces with insufficient population capacity and population pressure are mostly those in the western and central zones. In total, the provinces with excess population-carrying capacity and less population pressure were mainly in the central and eastern zones. And the provinces with insufficient population-carrying capacity and more population pressure are mainly in the western zone. To sum up, the eastern zone has excess population capacity and less population pressure, the central zone has a population-carrying capacity roughly equal to its total population, and the western zone has insufficient population-carrying capacity and more population pressure. This situation is in accord with the natural and economic conditions in these three zones. Therefore the existing distribution of population, dense in the east and sparse in the west, can be considered reasonable. Analysis of the relative deviation of population distribution in China also indicates that the 'rationality rate' of population distribution was more than 80 per cent (Wang Guixin, 1997). Accordingly, the concentration of population in the eastern zone of China since the economic reform in the late 1970s, and especially since the mid-1980s, indicates a positive transformation of China's population distribution leading to even greater rationality of distribution.

Prospects for Change in China's Population Distribution

Because of the influence of the stable regional structure of natural conditions including the topography and climate etc., the situation with regard to population distribution, dense in the east and sparse in the west, has been a long-standing feature in China. But regional disparities in level of economic development will continue to grow in China for some time to come as the economy develops under the market system. This is likely to cause continued concentration of the population in the eastern zone, at least until 2010.

Second, the birth rate, death rate and migration will show new trends, influencing population distribution in China. Death rates are comparatively low because of the improvement in medical, technical and welfare conditions, and the implementation of family-planning policy has also caused the fertility rate decline all over the country. As the regional differentials in birth and death rates are reduced, their influence on the population density becomes weak. With the continuation of family-planning policy and the improvement of medical, technical and welfare conditions, the natural increase of population determined by the fertility and death rates will have less and less effect on the population distribution in the next decades. Alongside natural population increase, migration has become a more active factor in determining the distribution of China's population since economic reform in the late 1970s. The main stream of people has been from the less developed central and western zones to the more developed eastern zone and has contributed substantially to the eastward-concentrating trend of population. Because the surplus rural labour force was numbered in millions, the reform of the household registration system, the weakening of the dual social structure and the relaxation of restrictions on population mobility made it inevitable that surplus labour would move. Furthermore, rapid economic development in the eastern zone and the enlargement of the regional disparity between the east and the west will cause population migration to become even more active in the future. People will continue to flow from the less developed western and central zones to the more developed eastern zone.

To sum up, future changes in China's population distribution will mainly be affected by economic development levels and migration, and will follow the tendency of eastward concentration that began in the mid-1980s at least until 2010, making distribution more rational. Deeply restricted by the stable physical environment structure, China's population distribution will not change significantly, and the basic model of China's population distribution, dense in the east, sparse in the west and mainly concentrated in southeastern sector, will continue for a long period.

Notes

1. Sichuan province was divided into two parts in 1997: Sichuan province and Chongqing municipality. In this chapter Sichuan province has its pre-1997 meaning and refers to a combination of these two.
2. The population data used here are mainly from the household registration system.
3. The division into three zones here was based on China's Seventh Five-Year Plan. The provinces in the eastern zone are: Beijing, Tianjin, Hebei, Liaoning, Shianghai, Jiangsu, Zhejiang, Fujian, Shandong, Guangdong, Guangxi and Hainan. Those in the central zone are: Shanxi, Neimenggu (Inner Mongolia),

Jilin, Heilongjiang, Anhui, Jiangxi, Henan, Hubei and Hunan. The western zone comprises: Sichuan, Guizhou, Shaanxi, Gansu, Qinghai, Ningxia, Xinjiang and Tibet (Xizang).

4. Here it refers to the GNP output per unit area.

References

Hu, Huanyong (1990), Division and prospect of population distribution in China, *Journal of Geography*, 2.

Wang, Guixin (1997), *Population distribution and regional economic development in China*. Shanghai: East China Normal University Press.

3 Trends and Regional Differentials in Fertility Transition

Tu Ping

In the past three decades, China has experienced a remarkable fertility transition from a level of about six children per woman down to less than two children per woman. Such an unprecedented transition in a very short period of time was, to a significant extent, initiated by a government-sponsored family-planning programme promoted under unfavourable social, economic and demographic conditions. It has enabled China to join the countries, mostly developed countries, that have achieved below-replacement fertility, and it will have profound social, economic and demographic implications for China.

There is no prevailing consensus about the relative importance of the major causes of fertility transition in China. However, the profound political, social and economic changes in the country have left a significant imprint on its demographic trends. This chapter examines the fertility trends in China since 1950, the current patterns and regional variations in fertility, and the possible social and demographic implications of a rapid fertility decline (see table 3.1).

Fertility Trends

The fertility transition in China can be roughly divided into six phases: a period of high fertility during the reconstruction period (1949–57); a big drop in the period of the Great Leap Forward (1958–61); a boom in the early 1960s; a rapid decline in the 1970s; a stagnation period in the 1980s; and the below-replacement fertility period after 1990 (see figures 3.1 and 3.2).

Initial high fertility period (1949–1957)

The return of peace after the foundation of the People's Republic in 1949 led to a period of high fertility with declining mortality. The crude birth rate for the country as a whole in this period was 32–8 per thousand of

Table 3.1 Changes in crude birth rates (CBR) and total fertility rate (TFR) by residence, China, 1950–1992

Year	CBR[a]			TFR[b]		
	Total	Urban	Rural	Total	Urban	Rural
1950	37.0	—	—	5.3	5.3	5.3
1951	37.8	—	—	5.3	5.1	5.3
1952	37.0	—	—	6.0	5.7	6.0
1953	37.0	—	—	5.7	5.5	5.7
1954	38.0	42.5	37.5	6.0	6.0	5.9
1955	32.6	40.7	31.7	6.0	5.7	6.1
1956	31.9	37.9	31.2	5.6	5.4	5.7
1957	34.0	44.5	32.8	6.2	6.2	6.2
1958	29.2	33.6	28.4	5.5	5.5	5.5
1959	24.8	29.4	23.8	4.2	4.4	4.2
1960	20.9	28.0	19.4	4.0	4.2	3.9
1961	18.0	21.6	17.0	3.3	3.1	3.3
1962	37.0	35.5	37.3	6.0	4.9	6.2
1963	43.4	44.5	43.2	7.4	6.3	7.6
1964	39.1	32.2	40.3	6.1	4.4	6.5
1965	37.9	26.6	39.5	6.0	3.8	6.5
1966	35.1	20.9	36.7	6.2	3.1	6.9
1967	34.0	—	—	5.3	2.9	5.8
1968	35.6	—	—	6.4	3.8	6.9
1969	34.1	—	—	5.7	3.3	6.2
1970	33.4	—	—	5.7	3.2	6.3
1971	30.7	21.3	31.9	5.4	2.8	6.0
1972	29.8	19.3	31.2	4.9	2.6	5.4
1973	27.9	17.4	29.4	4.5	2.4	5.0
1974	24.8	14.5	26.2	4.2	1.9	4.6
1975	23.0	14.7	24.2	3.6	1.8	4.0
1976	19.9	13.1	20.9	3.3	1.6	3.6
1977	18.9	13.4	19.7	2.9	1.6	3.1
1978	18.3	13.6	18.9	2.7	1.6	3.0
1979	17.8	13.7	18.4	2.8	1.4	3.2
1980	18.2	14.2	18.8	2.7	1.8	3.0
1981	20.9	16.5	21.6	2.5	1.5	2.8
1982	21.1	18.2	22.0	2.9	2.0	3.2
1983	18.2	16.0	19.9	2.6	1.8	2.8
1984	17.5	15.0	17.9	2.3	1.6	2.5
1985	17.8	14.0	19.2	2.3	1.5	2.6
1986	20.8	17.4	21.9	2.3	1.6	2.6
1987	21.0	17.6	22.2	2.6	1.8	2.8
1988	20.8	17.4	22.0	2.4	1.7	2.7
1989	20.8	16.0	22.3	2.3	1.6	2.5
1990	20.5	13.1	22.2	2.0	1.2	2.3
1991	19.3	13.6	20.6	1.9	1.2	2.0
1992	18.1	15.2	18.8	1.7	1.4	1.8

Sources:
[a] CPIRC (1990: 16); Jiang et al. (1995: 257–9).
[b] Coale and Chen (1987: 24–9); Feeney et al. (1993: 483, 486); Jiang et al. (1995: 222–4).

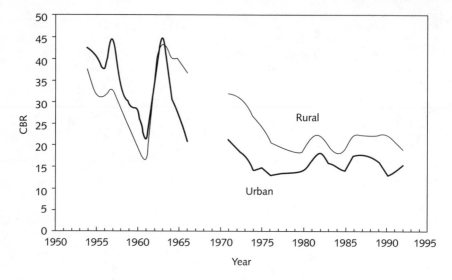

Figure 3.1   Changes in crude birth rate (CBR), 1950–1992

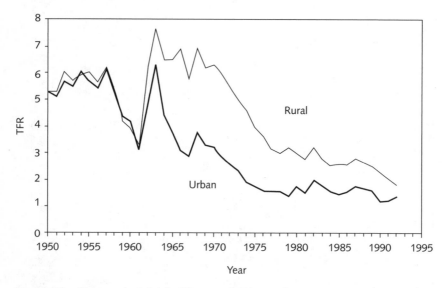

Figure 3.2   Changes in total fertility rate (TFR), 1950–1992

the population and the total fertility was around six children per woman.
There was little urban–rural differential in fertility (figure 3.2). Therefore,
the difference in birth rate between the urban and rural areas (figure 3.1)
was mainly caused by the difference in population age structure.

The Great Leap Forward (1958–1961)

During this period, mainly due to policy errors and a nationwide natural calamity, China experienced a large drop in fertility and a large number of excess deaths (Peng, 1987). The crude birth rate dropped from 34 in 1957 to only 18 in 1961 while the total fertility rate dropped from 6.2 to 3.3.

Post-famine recovery (1962–1970)

In this period, the birth rate and the total fertility rate reached a peak in 1963, as high as 43 and 7.4, respectively, because of compensatory child-bearing after the big drop in 1958–61. Fertility remained high thereafter except in urban areas where it started to decline very rapidly. Urban–rural differentials in fertility first appeared and increased rapidly in this period.

Rapid fertility decline (1971–1979)

In this period, the Chinese government introduced a national family-planning programme that promoted a policy of later birth, longer spacing, and fewer births (Wan, Xi, Shao). China's fertility transition was most remarkable in this period. The total fertility rate declined sharply from 6 in 1970 to 2.8 in 1979, an unprecedented drop of over three children per woman in less than nine years. Urban–rural differentials in fertility decreased to some extent, but remained fairly large throughout this period.

Stagnation period (1980–1989)

In spite of the one-child policy introduced in 1979 and various family-planning campaigns, fertility decline slowed down after the rapid decline observed in the 1970s. Both the crude birth rate and the total fertility rate experienced significant fluctuations during this period, around a level of 20 per thousand and 2.5 children per woman, respectively.

Below-replacement fertility period (1990–)

Fertility declined further and reached the below-replacement level (2.1) around 1991. Fertility in the urban and rural areas started to converge. However, the deterioration in the quality of demographic data in recent years has led to considerable disagreement and speculation about the recent fertility trend and its causes (Feeney and Yuan, 1994; Nygren and Hoem 1993; Zeng, 1995).

For a developing country such as China with a relatively low level of socio-economic development, the achievement in reducing fertility is remarkable. China is now far ahead of the rest of the developing world and very close to the developed countries in achieving low fertility. Although there is no prevailing consensus about the exact contribution of the main causes of China's fertility transition, the rapid fertility transition is generally attributed to: a strong and persistent commitment by the government to the control of rapid population growth, socio-economic development, and significant social and institutional changes (Peng and Tu, 1992).

Under the guidance of the central government, local governments at all levels have developed population plans that suit the local socio-economic conditions. Local resources have been mobilized and effective organizational arrangements, such as the family-planning-target responsibility system and the implementation of provincial family-planning regulations, have been made to meet the target of population control. The government's efforts in family planning have made an important contribution to success in birth control, although they have at the same time caused a great deal of concern among the international community.

At the same time, socio-economic development has also brought about a reduction in fertility. In the past four decades, especially after the economic reforms introduced in the late 1970s, there have been marked changes in the demand for children and the attitude towards childbearing. These changes have established a favourable context for the rapid decline in fertility.

China's demographic transition has benefited a great deal from the tremendous social reform in the last 40 years. The traditional familial role and function have been weakened and altered since the foundation of the People's Republic, and the costs and benefits of children have been changed too. Since 1949 China has achieved marked success in mass education. The illiteracy rate has dropped further in the recent years, from 22.8 per cent in 1982 to 15.9 per cent in 1990. This improvement in education has challenged the traditional value system, which hindered demographic transition. It has facilitated the diffusion of new ideological and behavioural norms, including those on birth control, and played an important role in China's fertility transition. All these changes have undermined the traditional pro-natal culture that put great importance on the continuation of the family line.

Fertility Patterns and Regional Differentials

The age patterns of fertility in China, by birth order, can be characterized as follows: early, universal and highly concentrated first births; relatively early and common second births; and low and late incidence of high-order (third and above) births (figure 3.3). Women give birth for the first time around the age of 22–3; over 90 per cent of women have had their first child before the age of 28, with an extremely low proportion remaining childless. Second birth is still fairly early, around age 25. According to the period parity progression analysis (table 3.2), about 99 per cent of the women in China will give birth at least once in their lifetime, three-

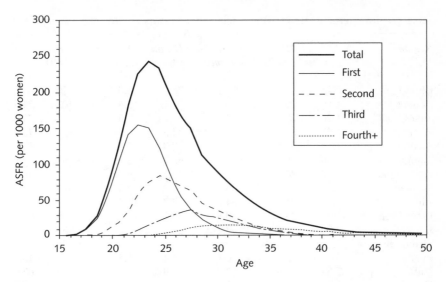

Figure 3.3   Age-specific fertility rate (ASFR) by birth order, 1989, whole China

Table 3.2 Period parity progression ratios (PPPR) by residence, China, 1989

| Residence | PPPR | | | |
	$0 \to 1$	$1 \to 2$	$2 \to 3$	$3 \to 4$
China	0.99	0.74	0.44	0.33
Cities	0.99	0.50	0.32	0.27
Towns	0.99	0.80	0.44	0.34
Counties	0.99	0.92	0.50	0.33

Source: Tu (1993a: 4).

quarters of the women who give birth once will go on to give birth a second time, but the progression ratio is much lower at higher parity. Parity progression ratios are highest in the counties (rural areas) and lowest in cities.

Significant regional differentials in fertility also exist in China. Table 3.3 shows significant variations in the total fertility rate (TFR) and mean age at childbearing among the 30 provinces in mainland China. The TFR is far below the replacement level in the two municipalities (Beijing and Shanghai) while it is still above 3 in provinces with a high proportion of minority populations (Guizhou, Xinjiang and Xizang). There is a strong association between fertility and the proportion of urban and minority

Table 3.3 Total fertility rate (TFR) and selected indicators by province, China, 1990

Region	TFR (1989)	MC (1989)	Population (millions)	Urban (%)	Minority (%)	GNP (RMB)	e(0)
China	2.25	26.1	1,115.1	26.2	8.0	1,410	69.3
Beijing	1.19	26.4	10.8	73.1	3.8	4,407	73.6
Shanghai	1.26	25.7	13.4	66.2	0.5	5,406	75.3
Zhejiang	1.43	25.1	40.5	32.8	0.5	1,839	72.5
Liaoning	1.54	25.5	39.6	50.9	15.6	2,455	70.9
Tianjin	1.60	26.0	8.7	68.7	2.3	3,361	72.8
Heilongjiang	1.71	24.9	34.3	47.2	5.7	1,576	68.2
Sichuan	1.79	25.3	105.2	20.3	4.6	923	67.2
Jilin	1.81	25.4	24.8	42.7	10.2	1,524	68.5
Inner Mongolia	1.91	25.7	20.8	36.1	19.4	1,188	67.0
Jiangsu	2.00	25.5	67.2	21.2	0.2	1,968	72.3
Shandong	2.11	27.0	82.4	27.3	0.6	1,458	71.3
Fujian	2.31	25.0	30.1	21.4	1.5	1,384	70.3
Hubei	2.39	25.8	53.8	28.9	4.0	1,338	67.6
Gansu	2.40	25.4	22.5	22.0	8.3	995	67.6
Hebei	2.40	26.4	59.4	19.1	3.9	1,289	71.8
Hunan	2.41	25.3	59.5	18.2	7.9	1,073	67.3
Shanxi	2.45	26.3	27.7	28.7	0.3	1,263	69.6
Jiangxi	2.49	24.9	37.6	20.4	0.3	1,002	66.8
Anhui	2.49	25.9	55.2	17.9	0.6	1,051	69.9
Yunnan	2.57	26.1	36.2	14.7	33.4	853	64.0
Guangdong	2.59	27.0	62.1	36.8	0.6	2,131	73.1
Qinghai	2.67	27.4	4.4	27.4	42.1	1,363	61.9
Shaanxi	2.69	26.4	31.9	21.5	0.5	1,075	68.4
Guangxi	2.71	26.8	41.9	15.1	39.1	847	69.3
Hainan	2.75	27.5	6.3	24.1	17.0	1,331	72.3
Henan	2.89	27.3	84.6	15.5	1.2	1,011	70.3
Ningxia	2.90	26.3	4.6	25.7	33.3	1,209	68.3
Guizhou	3.04	27.2	32.2	18.9	34.7	757	65.2
Xinjiang	3.04	28.4	15.1	31.9	62.4	1,498	65.1
Xizang	4.31	31.5	2.2	12.6	96.3	—	59.9

Sources: Tu (1993b: 7); Lu and Gao (1994: 55–6).
Note: MC is the mean age at childbearing.

populations at the aggregated level (table 3.3). Provinces with a high proportion of urban population and a low proportion of minority population
have achieved low fertility. It reflects the fact that deliberate fertility control
was first adopted by the Han population and started in urban and coastal
areas, then gradually adopted by other minority populations and diffused
to rural and inland areas. It also reflects differences in the family-planning
policy and its implementation between Han and minority populations and

Table 3.4 Selected indicators of the patterns of fertility, China, 1989

| Indicator | Fertility patterns | | | | | |
	I	II	III	IV	V	All
TFR	1.2	1.8	2.4	2.8	4.3	2.2
Mean age at childbearing	25.8	25.7	25.9	27.2	31.3	26.1
Mode age at childbearing	23	23	23	24	26	23
No. of provinces	2	9	13	5	1	30
Population (millions)	24.2	429.7	472.0	202.4	2.2	1,130.5
Urban population (%)	69.3	31.1	20.5	24.2	12.6	26.2
Minority population (%)	2.0	4.9	9.5	11.5	96.3	8.0

Source: Tu (1993b: 8).

between urban and rural areas. However, we should also note the considerable heterogeneity in degree of urbanization and ethnic composition of the population among provinces with a similar fertility.

A cluster analysis of the age-specific fertility schedules of the 30 provinces in China identifies five major patterns of fertility (table 3.4). Table 3.4 presents the estimated parameters of the fertility schedules along with a few demographic indicators for each cluster and for the country as whole. Cluster I consists of the 2 largest metropolitan cities with the lowest TFR (1.2–1.3) and has a population of 24.2 million (2.1 per cent of the total population); Cluster II consists of the 9 provinces with a TFR below the replacement level and has a population of 429.7 million (38.0 per cent). Clusters III and IV consist of 13 and 5 provinces with a TFR between 2 and 3 and which have populations of 472.0 million (41.8 per cent) and 202.4 million (17.9 per cent), respectively. And Cluster V consists of only Xizang (Tibet) which has the highest TFR (4.3) in the country and a population of 2.2 million (0.2 per cent).

The five clusters represent a wide range of fertility patterns at successive stages during the transition from high to extremely low fertility (see figure 3.4). The fertility schedule of Cluster V is characterized by a high overall level that spreads over a wide range of ages, representing a population with a weak deliberate fertility regulation. But the fertility schedule of Cluster I is characterized by an extremely low level of fertility that is heavily concentrated in a narrow age range, representing a population with extremely strong deliberate fertility regulation. Childbearing becomes more and more concentrated as the overall fertility level declines.

Concluding Remarks

In spite of the remarkable achievement in socio-economic development in the past four decades, China was still relatively backward in terms of

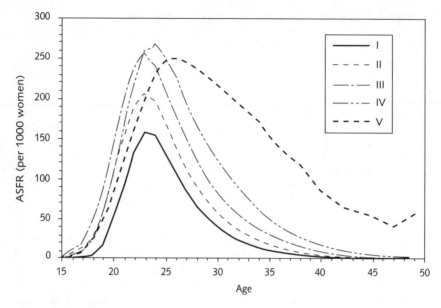

Figure 3.4   Age patterns of fertility, China, 1989

economic development when the fertility transition started. Therefore, it is expected that keeping China's fertility below the replacement level will require a continued effort in implementing family-planning policy, promoting socio-economic development, and developing and implementing other policies that encourage low fertility. While it is expected that China will continue its strong effort to control population growth and that the government sponsored family-planning programme will still have a strong influence on future fertility trends, whether China can successfully achieve its planned population goals in the future will also rely heavily on socio-economic development. The progress of China's political and economic reforms, as it leads to an increasing population mobility and changing government functions, will pose challenges to the existing family-planning programme. But at the same time, it will facilitate changes in people's lifestyle, fertility preferences and childbearing behaviour leading to fertility control on a voluntary basis. Therefore, economic development, modernization and the voluntary participation by the Chinese people in family planning will be the crucial factors determining future fertility trends in China.

With the unprecedented fertility decline to a level significantly below the replacement, rapid population ageing will be inevitable and China will face a severe conflict between the need to slow down its population growth and the need to avoid too rapid population ageing (Tu, 1995; Zeng and Vaupel, 1989). If the current fertility trend continues, the proportion of population

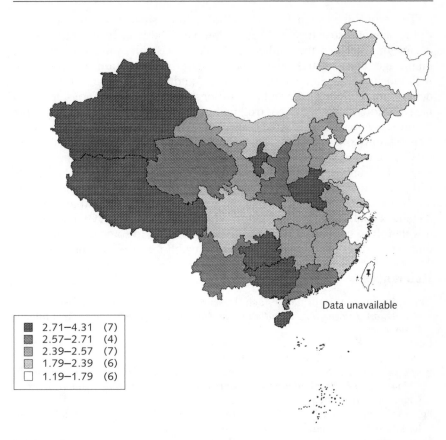

Data unavailable

■	2.71–4.31	(7)
■	2.57–2.71	(4)
▨	2.39–2.57	(7)
▨	1.79–2.39	(6)
☐	1.19–1.79	(6)

Figure 3.5   Total fertility rate by region, China, 1989

Source: As table 3.3.

aged 65 and over will increase from 5 per cent to 7 per cent in less than 20 years and from 7 per cent to 14 per cent in less than 26 years. For the proportion of population aged 65 and over to increase from 7 per cent to 14 per cent, it took France 115 years, Sweden 85 years, Germany and the United Kingdom 45 years, and Japan (the country with the most rapid pace of population ageing observed so far) 26 years. Therefore, China's population will age at a speed unprecedented in the world, if the currently predicted trends in fertility and mortality hold. Although it will in the short run experience a sharp decline in its dependency burden due to declining child dependency, it will eventually face an extremely high elderly and total dependency burden. It is very likely that China will face a severely aged population before it has had sufficient time and resources to establish an adequate social security and service system for the elderly (Tu, 1995).

The rapid decline in fertility will also have significant impact on the future size and structure of Chinese families. The impact of rapid fertility decline on their structure and dependency ratio is quite similar to its impact on the age structure and dependency ratio of the population as a whole, but with much greater fluctuations over time and an imbalance in the dependency burden among different generations (Tu, 1995). Therefore, China will face a serious challenge in controlling fertility but maintaining a reasonable family size and structure. Strict control of fertility will lead to a sharp reduction in family size and dependency burden in the short run, but it will eventually give the next generation an extremely heavy dependency burden. For single-child families, family kin such as sisters, brothers, uncles and aunts will simply disappear and the roles and positions of the remaining family kin will undergo significant changes. That will have profound and unpredictable social, economic and cultural implications (Zhou, 1996). Currently it is therefore of great importance to pay attention to some of the negative impacts of strict fertility controls and to search for effective solutions.

References

China Population Information and Research Centre (CPIRC), (1990), *China Population Information Handbook* (in Chinese). Beijing: Beijing Institute of Economics Press.

Coale, A. and Chen, S. L. (1987), *Basic Data on Fertility in the Provinces of China, 1940–82*. Honolulu: East-West Center.

Feeney, G., Luther, N. Y., Meng, Q. P. and Sun, Y. (1993), Recent fertility trends in China. *China 1990 Population Census: Papers for International Seminar*. Beijing: China Statistical Press.

Feeney, G. and Yuan, J. H. (1994), Below replacement fertility in China? A close look at recent evidence. *Population Studies* 48, 381–94.

Jiang, Z. H., Zhang, E. L. and Chen, S. L. (1995), *Statistics of 1992 Fertility Sample Survey in China* (in Chinese). Beijing: China Population Publishing House.

Lu, L. and Gao L. (1994), Abridged life tables by provinces in China in 1990. *Population Research* 18(3), 52–6 (in Chinese).

Nygren, O. and Hoem, B. (1993), Recent trends in fertility and contraceptive use in China: Results from a sample survey. Working Paper, Statistics Sweden.

Peng, X. Z. (1987), Demographic consequences of the Great Leap Forward. *Population and Development Review* 13 (4), 639–70.

Peng, X. Z. and Tu, P. (1992), Fertility and mortality transition in China. Paper presented at the Fourth Asian and Pacific Population Conference, 19–24 August 1992, Bali, Indonesia.

State Statistical Bureau (1991), *Ten Per Cent Sampling Tabulation on the 1990 Population Census of the People's Republic of China* (in Chinese). Beijing: China Statistical Press.

Tu, P. (1993a), Fertility patterns of Chinese women: Results based on parity progression analysis. *Population Science of China* 4, 1–5 (in Chinese).

Tu, P. (1993b), A comparative study of fertility patterns in China. Paper presented at the 22nd IUSSP General Population Conference, 24 August–1 September 1993, Montreal, Canada.

Tu, P. (1995), On the issues of population ageing and population control in China. *Population Sciences in China* 6, 61–70 (in Chinese).

Zeng, Y. (1995) Is China's fertility in 1991 and 1992 far below the replacement level? *Population Research* 19 (3), 7–14 (in Chinese).

Zeng, Y. and Vaupel, J. (1989), The impact of urbanization and delayed childbearing on population growth and ageing in China. *Population and Development Review* 15 (3), 425–46.

Zhou, Y. (1996), Fertility decline and kinship structure. *Population Science of China* 1, 37–41 (in Chinese).

4 Trends and Geographic Differentials in Mortality

Hao Hongsheng

Mortality Trends

For the period prior to 1949, China had an estimated crude death rate of about 25 per thousand population, an infant mortality rate of about 200 per thousand births, and a life expectancy at birth of 35 years (Ministry of Public Health (MPH), 1990: 69–75). These figures suggest that before the founding of the People's Republic, China was a country with typically high mortality rates. During the past few decades, China has experienced a very rapid mortality transition process and become a country of low mortality. Today China's mortality level is among the lowest in the developing world.

As a rough indicator of mortality levels, the trend of decline in the crude death rate (CDR) is shown in table 4.1. It can be seen from the table that the most striking decline in annual death rates took place in the 1950s. From 1949 to 1957, the CDR was reduced by almost half. The process of decline was then interrupted by the Great Leap Forward in 1958 and the following three-year famine period. The annual death rates started to increase in 1958, reached a peak in 1960, and returned to normal in 1962. After the famine, the CDR continued to decline at a fairly fast pace and reached a low level of below seven per thousand by the end of the 1970s. From then on, the decline of CDR came to a halt and was succeeded by a fluctuation between six and seven per thousand in the 1980s and the 1990s. Such a change in CDR does not mean the mortality conditions in China stopped improving, instead it has been, to a large extent, an effect of changing age composition, that is to say that there have been more older people in the population. This change in age composition has offset the effect of the improvement in mortality conditions on CDR and hindered the CDR from further decline. For later years, therefore, the function of the CDR in depicting the trends in mortality tends to diminish. With the rapid ageing process that China's population is undergoing, we may even expect a moderate increase in China's CDR in the years to come.

Infant mortality rate (IMR) is not only an important indicator of the health and mortality situation, but is also widely considered as an indicator

Table 4.1 Crude death rates, 1949–1990 (deaths per 1,000 people)

Year	CDR	Year	CDR	Year	CDR
1949	20.00	1964	11.50	1979	6.21
1950	18.00	1965	9.50	1980	6.34
1951	17.80	1966	8.83	1981	6.36
1952	17.00	1967	8.43	1982	6.60
1953	14.00	1968	8.21	1983	6.90
1954	13.18	1969	8.03	1984	6.82
1955	12.28	1970	7.60	1985	6.78
1956	11.40	1971	7.32	1986	6.86
1957	10.80	1972	7.61	1987	6.72
1958	11.98	1973	7.04	1988	6.64
1959	14.59	1974	7.34	1989	6.54
1960	25.43	1975	7.32	1990	6.67
1961	14.24	1976	7.25	1991	6.70
1962	10.02	1977	6.87	1992	6.64
1963	10.04	1978	6.25		

Source: Yao and Yin, 1994: 9.

Table 4.2 Infant mortality rate, 1944–1987 (per 1,000 live births)

Year	IMR	Year	IMR	Year	IMR
1944–9	203.60	1970	51.95	1983	41.37
1950	197.93	1975	48.05	1984	38.41
1955	107.64	1980	42.76	1985	37.38
1960	109.92	1981	37.33	1986	37.14
1965	72.13	1982	36.42	1987	39.92

Source: Yao and Yin, 1994: 145.

of development, since it is very sensitive to socio-economic changes. The 1990 population census reported an infant mortality rate of 32.9 : 32.4 for males and 33.5 for females (State Statistics Bureau (SSB), 1995). Table 4.2 presents the IMR estimates for China since 1944. The table shows that China's infant mortality has dropped tremendously over the last four decades, and that the greatest reduction in IMR happened in the 1950s and 1960s. By the early 1970s, IMR was only about one-fourth of the level before 1949. During the 1980s and 1990s IMR continued to decline, but at a much slower rate.

Underreporting infant mortality has long been a worldwide phe-nomenon, and China is no exception. Alternative data sources for IMR estimation provide evidence that the infant mortality rates have been under reported. A survey conducted under the guidance of the Ministry of Public Health in 1986 found that, on average, the IMRs for 1986 at survey localities were substantially higher than those reported by the 1982 census, especially in the rural areas. Based on the survey findings and certain assumptions, the IMR for 1986 was estimated as 51.1 per thousand for China (Zhou et al., 1989). A more recent survey, the 1991–3 National Sample Survey on Child Mortality conducted by the Ministry of Public Health, reported that the infant mortality was 50.2 for 1991, 46.7 for 1992, and 42.7 for 1993 (Lin et al., 1996). The infant mortality data from various sources undoubtedly indicate a rapid decline in IMR over the past four decades. The exact current level of IMR, however, remains an issue to be further investigated.

In the light of the declining mortality trends shown above, a great in-crease in life expectancy at birth in China can also be expected. Table 4.3 presents the reported life expectancy for selected years from different data sources. From an estimated figure of about 35 years before 1949, life expectancy had increased to nearly 70 years in 1990, a very rapid increase compared to other developing countries in the post-war era.

International comparison of life expectancy in table 4.4 shows that although China is still a low-income country, China's mortality level is among the lowest in the less developed world. China's life expectancy is not only higher than that of the low-income countries, but also higher than the world average for middle-income countries.

The substantial reduction in mortality in the most populous country in the world within a few decades, especially during the 1950s, has been a

Table 4.3 Reported life expectancy at birth for selected years (in years)

Year	Data source	Total	Males	Females
1973–5[a]	Cancer Epidemiology Survey	—	63.62	66.31
1981[b]	1982 National Population Census	67.88	66.43	69.35
1986[c]	1987 National Population Sample Survey	68.90	67.00	71.00
1985–8[d]	1988 Fertility and Birth Control Sample Survey	69.24	67.65	70.94
1989[e]	1990 National Population Census	68.55	66.84	70.47

Sources: [a] Rong et al. 1981: 51–2.
[b] *Yearbook of China's Population 1985* (CASS, 1986: 880–9).
[c] Rao and Chen, 1991: 59.
[d] Yan and Chen, 1990: 21–3.
[e] SSB, 1995.

Table 4.4 Comparison of China's life expectancy with countries of different income levels, 1991

Income level	GNP per capita (dollars)	Life expectancy at birth (years)
Low income	350	62
China	370	69
Middle income	2,480	68
Lower middle income	1,590	67
Upper middle income	3,530	69
Low and middle-income	1,010	64
High income	21,050	77
World	4,010	66

Source: The World Bank (1993), *World Development Report 1993*, Oxford University Press, 238–9.

remarkable achievement by any standards. There are many possible explanations for it.

First, the first half of the twentieth century in China's history was characterized by continuous wars, revolutions and famines. The social and political changes in 1949 ended many years of war and chaos and re-established stability in society.

Second, the standard of living and the food supply for the majority of the population improved both because of economic development and, particularly, because of a more egalitarian resource distribution. This distribution system was intended not to encourage individuals to amass wealth, but to ensure that everyone had a share of the basic necessities of life.

Third, the Chinese government adopted well-devised health policies in the early 1950s. Among the four major policies, the policy of 'giving priority to prevention' has played a very important role in the reduction of sickness and mortality, and proved particularly successful.[1]

Fourth, China established a national network of health services, which has been expanding rapidly over the past 40 years. In 1949 there were only 80,000 hospital beds and 541,240 medical personnel nationwide, mostly concentrated in the cities. By 1990 these numbers had increased to 2,624,086 and 4,906,201, respectively, with a large proportion in the rural areas (MPH, 1990: 4–6). Particular noteworthy is the three-level (county, township and village) medical and health network in rural China, which has been providing peasants with cheap but effective means of prevention and treatment. The medical institutions at the three levels co-operate with each other, and jointly undertake the tasks of treatment, prevention, maternity and child health care, health education and medical training (*China's Public Health Yearbook 1983*: 190–3).

Geographic Differentials in Mortality

China is a large country not only in terms of population, but also in terms of geographic size. As a result, there are vast differences in natural environment, culture and socio-economic development between urban and rural areas and among the various regions of China. Such differences are reflected in the health and mortality situation. It is therefore necessary to study geographic differentials, in order to get a better understanding of the mortality situation in the country as a whole.

Urban–rural mortality differentials

Table 4.5 presents the trends in China's urban–rural differentials in the reported crude death rates. Since the 1950s cities and counties have taken a similar path of rapid mortality decline, but the county CDRs have always been higher than those of cities. During the 1950s and 1960s the difference even widened because of a faster decline in city death rates. The figures for 1959–61 show that the impact of the famine was much worse in the rural areas than in the cities. In 1960, when famine mortality was at the maximum, the reported city CDR increased roughly by half, while the reported county CDR more than doubled as compared to normal levels before the famine.

Table 4.5 Crude death rates by urban and rural residence (deaths per 1,000 population)

Year	Urban	Rural	Year	Urban	Rural
1954	8.07	13.71	1975	5.39	7.59
1955	9.30	12.60	1976	6.60	7.35
1956	7.43	11.84	1977	5.51	7.06
1957	8.47	11.07	1978	5.12	6.42
1958	9.22	12.50	1979	5.07	6.39
1959	10.92	14.61	1980	5.48	6.47
1960	13.71	25.58	1981	5.14	6.53
1961	11.61	14.89	1982	5.28	7.00
1962	8.28	10.32	1983	5.92	7.69
1963	7.13	10.49	1984	5.86	6.73
1964	7.27	12.17	1985	5.96	6.66
1965	5.69	10.06	1986	5.75	6.74
1966	5.59	9.47	1989	5.78	6.81
1971	5.35	7.57	1990	5.71	7.01
1972	5.29	7.93	1991	5.50	7.13
1973	4.96	7.33	1992	5.77	6.91
1974	5.24	7.63			

Source: Yao and Yin, 1994: 8.

During the early 1970s the city CDRs were fairly stable at a low level, while at the same time the county CDRs continued to decline at a moderate speed. As a result, the city and county CDRs started to converge slightly. This narrowing of the gap between city and county CDRs was partially due to the effect of a faster decline in mortality in the rural areas during the period, and partially due to a simultaneous effect of differential age composition on CDRs, because higher rural fertility caused an age composition younger than that in the urban areas.

Table 4.6 presents the IMR estimates by city and county based on a retrospective national fertility survey in 1988. It shows that in both areas IMR had been decreasing fast, but a gap apparently still existed between city and county. Before 1949 IMRs for both the city and county were very high, but the gap was not very large. In the 1950s the city IMR declined faster and the city–county difference increased. During the 1960s the difference began to converge, but from the 1970s to the 1980s, the relative difference between the city and county IMRs barely narrowed. Other sources reported much larger urban–rural differences in infant mortality. The two surveys by the Ministry of Public Health estimated that the IMRs, for urban and for rural areas respectively, were 20.0 and 59.3 per thousand for 1986 (Zhou et al., 1989), and 17.3 and 58.0 per thousand for 1991 (Lin et al., 1996).

As would be expected, the urban–rural difference in life expectancy has also been sizeable. Table 4.7 presents life expectancy for city, town and county based on the 1982 and 1990 censuses. One can see that there has been a gap of 3 to 4 years between urban (city and town) and rural areas (county).

The urban–rural difference has created a fundamental distinction in China's society. Differences persist in macro aspects such as socio-economic

Table 4.6 Estimates of infant mortality by city and county, 1945–1949 to 1985–1988 (deaths per 1,000 live births)

Year	Infant mortality rate		County/City (%)
	City	County	
1945–49	181.09	206.97	114
1955–59	74.82	130.54	174
1965–69	42.79	67.24	157
1975–79	31.41	45.52	145
1980–84	29.18	42.02	144
1985–88	26.96	38.76	144

Source: Yan and Chen, 1990.

Table 4.7 Life expectancy by residence, 1981 and 1989–1990

Residence	1981			1989–1990		
	Males	Females	Both sexes	Males	Females	Both sexes
City	69.08	72.74	70.87	70.70	75.05	72.77
County	65.56	68.36	66.95	67.59	70.91	69.18

Source: Hunag and Liu, 1995: 27–56.

Table 4.8 Number of hospital beds and medical personnel per 1,000 population by city and county, 1949–1990

Year	Hospital beds			Medical personnel		
	Nation	City	County	Nation	City	County
1949	0.15	0.63	0.05	0.93	1.87	0.73
1957	0.46	2.08	0.14	1.61	3.60	1.22
1965	1.06	3.78	0.51	2.11	5.37	1.46
1975	1.74	4.61	1.23	2.24	6.92	1.41
1978	1.94	4.85	1.41	2.57	7.73	1.63
1980	2.02	4.70	1.48	2.85	8.03	1.81
1985	2.14	4.54	1.53	3.28	7.92	2.09
1990	2.32	4.18	1.55	3.45	6.59	2.15

Source: China's Health Statistics Digest 1990: 7.

development, environment, living standards, education, health care and life-style between urban and rural areas in China. These differences can lead to mortality differentials through many factors including food and nutrition, water, housing conditions, child-care practices, access to medical facilities, which are all closely related to morbidity and mortality.

A fundamental difference in health conditions between urban and rural areas has been the relative provision of medical services. Since 1949 the focal point of the government's health policy has been to stress the development of the health-care system in the rural areas. In fact, rural medical services, in the form of a 'three-level medical and health network', have been growing fast and have played an important role in China's rapid mortality decline. Nevertheless, large differences in medical services remain between urban and rural areas. As table 4.8 shows, there has been substantial development of medical services for both urban and rural areas, as indicated by the number of hospital beds and medical personnel during the past four decades, but the large city–county differences are clear from the figures.

It is important to point out that the data in this table does not fully reflect the broad urban–rural differences in the quality of the service. Such differences largely lie in the qualification levels of medical personnel, which have been substantially lower in rural areas than in towns and cities. In the early 1980s the ratio of medical personnel with senior, middle and lower levels of qualifications was reported to be 1 : 1.2 : 0.7 for the urban areas, and 1 : 2.7 : 2 for the rural areas. Furthermore, many lower level personnel in the rural areas were poorly trained or had no medical training (*Yearbook of China's Public Health 1983*, 1984: 192). In the late 1980s it was reported that most of the 1.27 million medical personnel in the 738,000 villages of China had never received mid-level medical training (*Jiankang Bao* [Health News Daily], 15/11, 1988).

Another factor is the water supply, an important determinant of exposure to disease, especially diarrhoeal diseases. In China there is still a considerable urban–rural difference in the availability of safe drinking water. Most urban residents have a supply of tap water, but by the end of 1989 only 26 per cent of rural residents had access to it (MPH, 1990: 86).

Regional mortality differentials

Differences in socio-economic and environmental levels among the various regions of China are even more pronounced than the urban–rural differences. While some coastal regions have been experiencing a rapid process of economic growth and industrialization, a few northwest and southwest mountainous provinces, characterized by a harsh natural environment and a high proportion of ethnic minorities, have lagged far behind in socio-economic development and have not been completely lifted out of poverty and backwardness. One would, consequently, expect to see large mortality differentials between regions.

Listed in table 4.9 are regional CDRs for selected years which show the general trends in regional mortality differentials during the last four decades. From the table we can see that virtually all regions have experienced a dramatic decline in mortality,[2] and that the differences in the CDRs among regions has been narrowing over time.

In the 1950s, the regions with the highest death rates had a reported CDR at least twice as high as that of regions at the lowest rates. The largest-ever regional differences in CDRs occurred during the famine year of 1960, when the highest reported CDR, recorded in Anhui, was 10 times higher than that of Shanghai, which had the lowest rate that year. During the 1970s and the 1980s, with the decline in the general levels of mortality, differences among regional CDRs showed a converging trend, partially due to the effect of differential age composition. However, the regional disparities in CDR were still noticeable.

Table 4.9 Regional crude death rates for selected years (deaths per 1,000 population), 1950–1990

Region	1950	1955	1960	1965	1970	1975	1980	1985	1989–1990
Beijing	14.60[a]	9.60	9.22	6.70	6.37	6.53	6.30	5.50	5.43
Tianjin	*11.12*	*8.68*	*7.46*	*5.26*	*6.26*	*6.59*	6.03	5.78	5.98
Hebei	*12.38*	*11.86*	*12.19*	8.38	*6.52*	7.22	7.27	5.69	5.76
Shanxi	*13.51*	*12.93*	*14.21*	10.40	8.13	7.85	6.73	5.98	6.25
Inner Mongolia	—	*11.46*	*9.70*	*9.59*	5.81	*6.06*	5.46	4.46	5.79
Liaoning	10.80[a]	*9.40*	*11.50*	7.10	*5.10*	*6.10*	5.60	5.27	6.01
Jilin	*12.40*	*9.91*	*10.13*	9.70	*6.37*	*6.71*	6.20	5.34	6.12
Heilongjiang	—	*11.33*	*10.52*	*8.00*	5.81	5.43	7.24	4.29	5.33
Shanghai	*7.70*	8.21	*6.90*	5.70	*5.00*	6.01	6.49	6.69	6.36
Jiangsu	—	*11.65*	*18.41*	9.50	*6.85*	6.46	6.57	5.87	6.07
Zhejiang	*15.43*	*12.58*	*11.88*	8.10	*5.96*	6.31	6.29	6.05	6.10
Anhui	*7.05*	*11.80*	68.58	*7.24*	*6.45*	*5.68*	4.73	5.16	5.79
Fujian	*13.92*	*8.26*	*15.61*	7.30	*6.04*	6.54	6.54	5.38	5.70
Jiangxi	—	*16.23*	*16.06*	*9.39*	*8.15*	8.01	5.30	5.54	6.59
Shandong	12.20[a]	*13.73*	23.51	10.20	*7.34*	7.53	6.61	5.90	6.25
Henan	—	*11.75*	*39.56*	*8.45*	*7.61*	*7.66*	5.22	6.22	6.18
Hubei	—	*11.60*	*21.22*	10.00	7.70	*7.88*	7.09	6.69	6.84
Hunan	—	16.41	29.26	11.20	10.16	8.34	6.87	6.46	7.07
Guangdong	*14.05*	*10.70*	*15.09*	*6.82*	*5.85*	*6.06*	5.41	5.01	5.34
Guangxi	—	*14.80*	*29.20*	9.00	*6.77*	*6.77*	5.88	5.05	5.96
Hainan	—	—	—	—	—	—	—	6.67	5.22
Sichuan	—	13.16	*47.78*	11.50	12.60	8.86	6.81	6.67	7.06
Guizhou	*16.61*	*16.24*	*52.33*	*15.16*	*10.84*	*10.53*	7.04	6.37	7.13
Yunnan	*17.86*	*13.76*	*26.26*	*12.97*	*8.15*	*8.68*	7.42	6.56	7.71
Tibet	—	—	—	—	7.64	9.05	8.20	10.13	9.20
Shaanxi	—	*10.55*	*12.27*	13.00	*6.82*	8.16	7.23	5.99	6.49
Gansu	*11.00*	*11.89*	*41.32*	12.30	*7.92*	7.42	5.53	4.99	5.92
Qinghai	—	13.76	40.73	9.10	7.56	8.24	6.09	4.58	6.84
Ningxia	—	—	13.88	9.30	6.43	7.74	5.67	3.77	5.07
Xinjiang	*19.92*	*14.40*	*15.67*	11.10	*8.17*	8.74	7.70	6.39	6.39

Sources: Figures for 1950 to 1985 from *A Collection of Population Statistical Data of PRC, 1949–1985* (SSB and MPS, 1988: 273–399). 1990 figures from *Major Figures from the Fourth National Population Census of the PRC* (PCO, 1991: 16), for the period from 1 July 1989 to 30 June 1990.
Notes: Figures in *italics* are officially adjusted death rates.
[a] From *Yearbook of China's Population 1985* (CASS, 1986: 399, 429, 491).

The regional infant mortality rates based on the 1990 censuses are presented in table 4.10. The large variations in regional IMRs are apparent. While the IMRs of the more developed regions were already well below 20 per thousand live births, the less developed regions in the western part of the country had an IMR of over 70 per thousand. In Tibet and Xinjiang, the 1990 IMR was even over 90 per thousand. The regional variations in IMRs can also be clearly viewed from a map, as figure 4.1 shows.

Table 4.10 Infant mortality rate by region, 1990 (per 1,000 live births)

	Region	Male	Female	Both sexes
	Total	32.4	33.5	32.9
1	Beijing	12.2	10.7	11.5
2	Tianjin	15.1	13.6	14.4
3	Hebei	17.4	15.4	16.5
4	Shanxi	24.8	23.9	24.4
5	Inner Mongolia	39.7	37.2	38.5
6	Liaoning	21.4	18.8	20.2
7	Jilin	26.2	23.3	24.8
8	Heilongjiang	34.2	27.2	30.8
9	Shanghai	15.1	12.2	13.7
10	Jiangsu	21.2	22.4	21.8
11	Zhejiang	22.2	25.7	23.8
12	Anhui	26.6	29.7	28.1
13	Fujian	23.5	27.9	25.6
14	Jiangxi	42.2	52.2	46.9
15	Shandong	16.1	17.6	16.8
16	Henan	19.0	22.2	20.5
17	Hubei	31.5	30.2	30.9
18	Hunan	43.8	44.2	44.0
19	Guangdong	17.0	17.8	17.4
20	Guangxi	33.1	55.4	43.2
21	Hainan	37.7	36.1	37.0
22	Sichuan	45.5	47.4	46.4
23	Guizhou	63.0	59.1	61.0
24	Yunnan	74.6	66.1	70.5
25	Tibet	102.6	84.0	93.4
26	Shaanxi	31.7	30.8	31.3
27	Gansu	32.2	32.3	32.3
28	Qinghai	91.0	76.7	84.1
29	Ningxia	52.6	42.1	47.5
30	Xinjiang	100.2	84.6	92.6

Source: SSB, 1995.

Similarly, regional differentials in mortality can also be observed in terms of life expectancy. Even though life expectancy for China as a whole has been substantially higher than the average level in developing countries, the comparison of regional life expectancy based on the 1990 census revealed the magnitude of internal inequality in mortality among regions. Table 4.11 shows that in 1990, the highest regional life expectancies, which were observed in Shanghai, were about 73 years for males, 77 years for

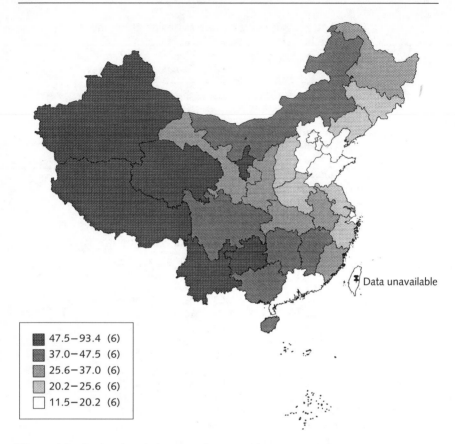

Data unavailable

47.5−93.4 (6)
37.0−47.5 (6)
25.6−37.0 (6)
20.2−25.6 (6)
11.5−20.2 (6)

Figure 4.1   Regional variations in infant mortality rates, 1990

females, already close to the average life expectancy for high-income countries during the same period. At the other extreme, the lowest regional life expectancies observed, in Tibet, were about 58 and 62 for males and females respectively, below the average level for low-income countries. The gap between the two extremes was around 15 years for both sexes. The actual differences in mortality would be even more pronounced if lower levels of units were used in the analysis, since a large province in China is larger than most countries in the world, and contains enormous internal differences.

Figures 4.2 and 4.3 show the regional patterns of mortality level indicated by life expectancy. It can be seen clearly from the maps that the regional mortality patterns are highly consistent with the regional pattern of socio-economic development. The regions with low mortality levels are mainly those coastal provinces in the east with higher development levels, while the high-mortality regions are the ones located in the northwest and

Table 4.11 Life expectancy at birth by region, 1990 (years)

	Region	Male	Female	Both sexes	Sex difference (female–male)
	Total	66.84	70.47	68.55	3.63
1	Beijing	71.07	74.93	72.86	3.86
2	Tianjin	71.03	73.73	72.32	2.70
3	Hebei	68.47	72.53	70.35	4.06
4	Shanxi	67.33	70.93	68.97	3.60
5	Inner Mongolia	64.47	67.22	65.68	2.75
6	Liaoning	68.72	71.94	70.22	3.22
7	Jilin	66.65	69.49	67.95	2.84
8	Heilongjiang	65.50	68.73	66.97	3.23
9	Shanghai	72.77	77.02	74.90	4.25
10	Jiangsu	69.26	73.57	71.37	4.31
11	Zhejiang	69.66	74.24	71.78	4.58
12	Anhui	67.75	71.36	69.48	3.61
13	Fujian	66.49	70.93	68.57	4.44
14	Jiangxi	64.87	67.49	66.11	2.62
15	Shandong	68.64	72.67	70.57	4.03
16	Henan	67.96	72.55	70.15	4.59
17	Hubei	65.51	69.23	67.25	3.72
18	Hunan	65.41	68.70	66.93	3.29
19	Guangdong	69.71	75.43	72.52	5.72
20	Guangxi	67.17	70.34	68.72	3.17
21	Hainan	66.93	73.28	70.01	6.35
22	Sichuan	65.06	67.70	66.33	2.64
23	Guizhou	63.04	65.63	64.29	2.59
24	Yunnan	62.08	64.98	63.49	2.90
25	Tibet	57.64	61.57	59.64	3.93
26	Shaanxi	66.23	68.79	67.49	2.56
27	Gansu	66.35	68.25	67.24	1.90
28	Qinghai	59.29	61.96	60.57	2.67
29	Ningxia	65.95	68.05	66.94	2.10
30	Xinjiang	61.95	63.26	62.59	1.31

Source: SSB, 1995.

southwest and characterized by backward socio-economic development and high proportions of minority nationalities. The regions with mid-level mortality are mostly those located between the coastal and west regions with mid-level development.

Regional sex pattern of mortality

As the general mortality level of China decreases, the sex differential in mortality has been widening. According to the 1990 census data, women in

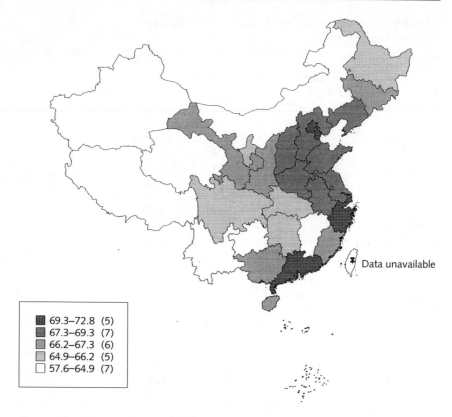

Figure 4.2   Life expectancy at birth, male, 1990

China tend to live about three and half years longer than men, while the female advantage was less than three years at the time of the 1982 census. An analysis based on China's past census data revealed that with each one-year increase in life expectancy for both sexes combined, the female advantage in life expectancy tends to expand by 0.21 years (Hao, 1995).

Since the sex differential in mortality is related to general level of mortality, it is no surprise to see large regional variations in sex differentials of mortality, indicated by gaps between the life expectancies of men and women. Table 4.11 shows that although life expectancy for women is higher than for men in all regions in China, female advantage in life expectancy has been much more apparent in some regions. The largest sex gaps in life expectancy, found in Hainan and Guangdong, were as much as five or six years, while in Xinjiang and Gansu, the female advantage was less than two years. There seems to be a link between the sex gap in mortality and the overall level of mortality. However, such a relationship has not been definitely established.

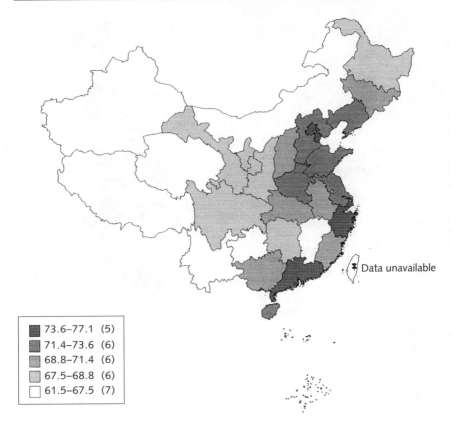

■	73.6–77.1 (5)
■	71.4–73.6 (6)
▦	68.8–71.4 (6)
▨	67.5–68.8 (6)
□	61.5–67.5 (7)

Data unavailable

Figure 4.3 Life expectancy at birth, female, 1990

Some degree of geographic homogeneity is detectable in figure 4.4. Regions with the largest sex gap in life expectancy are concentrated along the southeast coast. The sex gap tends to be modest in the north-central regions, and smaller in the northeast and the south-central regions. The smallest female advantage in life expectancy was observed in the northwest regions of China. In general, regions in north China tend to have a smaller female advantage in life expectancy than in the south, when levels of mortality are controlled for. The pattern of geographic homogeneity with regard to sex differentials in mortality observed in the 1990 census, was consistent with what was found in the previous census.

Summary

Various indicators have clearly shown a rapidly declining trend in mortality in China. During the past few decades, life expectancy has doubled, and

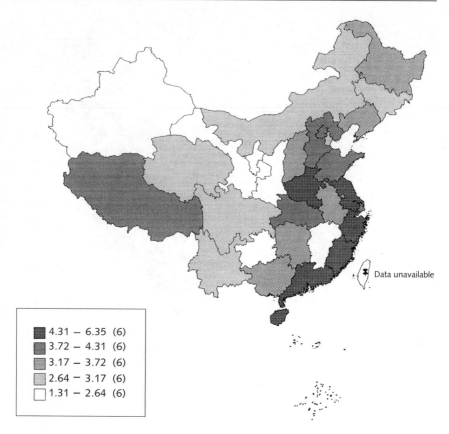

Figure 4.4 Sex difference in life expectancy, female–male, 1990

infant mortality has been reduced to a fraction of what it was in 1949. Without doubt, socio-economic development has been responsible for these achievements, but such a dramatic change in this huge population has also reflected the success of China's preventive health policy and the effectiveness of the health care system.

Although China's overall level of mortality has decreased greatly, inequality in health conditions and longevity remains. Recent census data revealed a gap of 3 to 4 years in life expectancy between the urban and rural areas. The regional gap was even larger: about 15 years difference between the regions at the two opposite extremes of the spectrum. Gaps of such magnitude constitute a great challenge to China's health sector, as well as a great potential for the further improvement of Chinese longevity. Analysis by region has detected a geographic pattern in mortality levels.

With the decline in the overall level of mortality, differential changes have also altered the sex patterns of mortality. As the overall level of life expectancy has increased, the gap in life expectancy between the two sexes has been widening. Analysis based on regional data also shows a geographic pattern in the sex differentials in mortality.

Notes

1. The four major health policies are: 'gearing to the needs of workers, peasants and soldiers'; 'giving priority to prevention'; 'consolidation between traditional and western medicine'; 'combining public health work with the mass movement'. The first three were adopted during the First National Conference on Public Health in 1950, and the fourth was added by Premier Zhou Enlai during the Second National Conference on Public Health in 1952 (Qian 1983: 10–12).
2. The figures for Tibet seem too low, especially for the years before the 1980s, and therefore may not reflect the real trend in mortality.

References

China's Public Health Yearbook 1983 (in Chinese) (1984), Beijing: People's Public Health Press.

Department of Population Statistics, State Statistical Bureau (SSB) and Ministry of Public Security (MPS) (1988), *A Collection of Population Statistical Data of the PRC* (in Chinese). Beijing: China Finance and Economy Press.

Hao, Hongsheng (1995), A study on sex differential of mortality in China. *Population Science of China*, 47, 2–11 (in Chinese).

Huang, Rongqing and Liu, Yan (eds) (1995), *Mortality Data of China's Population* (in Chinese). Beijing: China Population Publishing House.

Lin, Liangming et al. (1996), Levels and trends in infant mortality in China, 1991–1993. *Population Research*, 20 (4), 50–6 (in Chinese).

Ministry of Public Health, PRC (1990), *China's Health Statistical Digest 1990* (in Chinese). Beijing: printed by The Ministry of Public Health.

Population Census Office, PRC (PCO) (1991), *Major Figures from the Fourth National Population Census of China* (in Chinese). Beijing: China Statistical Publishing House.

Population Research Institute, China Academy of Social Sciences (CASS) (1986), *Yearbook of China's Population 1985* (in Chinese). Beijing: China Social Sciences Publishing House.

Qian, Xinzhong (1984), A Review of the victorious development in the cause of public health in our country. *China's Public Health Yearbook 1983* (in Chinese). Beijing: People's Public Health Press.

State Statistics Bureau (SSB) (1995), Regional infant mortality rates and life expectancy in China, 1990. *Market and Demographic Analysis*, 1, 4–62 (in Chinese).

Yan, Rui and Chen, Shengli (1990), An analysis of age-specific death rates and levels of longevity of China's population over the past forty years (in Chinese). Paper presented at the Seminar on the National Fertility and Birth Control Sample Survey, November 1990, Hangzhou.

Yao, Xinwu and Yin, Hua (eds) (1994), *Basic Data of China's Population*. Beijing: China Population Publishing House.

Zhou, Youshang et al. (1989), An analysis of China's infant mortality. *Population Science of China*, 3, 35–46 (in Chinese).

5 Population Policy and the Family-Planning Programme

Xie Zhenming

China did not experience a population explosion until the year 1700. Since then, the total population of China has increased dramatically from 175 million in 1700 to 1.2 billion in 1995, and it is very likely to reach 1.3 billion by the year 2000. That is seven times the size of 300 years ago. The fastest growth took place during the last 50 years of the twentieth century, when the founding of the People's Republic in 1949 brought a population of 540 million peace and stability after decades of war. The total population had reached 800 million in 1969, before the national family-planning programme started, with an annual growth rate of more than 2 per cent.

In recent decades, rapid socio-economic development has resulted in an increase in population awareness amongst Chinese leaders. China decided to join the trend in international population activities by promoting family planning among its population and, as a result, the family-planning programme was expanded to cover almost the whole country, with a strong commitment from the Chinese government, in the early 1970s. The achievements of fertility control in China are well known; the crude birth rate dropped from 33.43 per thousand in 1970 to 17.12 per thousand in 1995; the natural growth rate decreased from 2.58 per cent to 1.06 per cent; and the total fertility rate declined from 5.81 to 2.00 during the same period (Yao and Yin, 1994; Peng, 1995). It is roughly estimated that the total number of births averted due to family planning in China over the past 25 years may be more than 250 million. To put it in other way, it would add another 100 million to China's population if there had been no family-planning programme in the past decade. The family-planning programme has exerted a strong and positive impact on socio-economic development and has, therefore, made a considerable contribution to the stability of the world population.

Population Policy

China's population policy is well known and its explicitly stated aim is 'controlling population quantity and improving the quality of life'. In general, the population policy focuses on controlling the number of births, and improving the quality of life mainly by achieving better physical health and education. The premise is that the higher the quality of the 'producers', the better the human resources become that are available for economic development. But when the rapidly increasing population becomes a burden, the priority given to improving the quality of life may be neglected. At the Fifth National People's Congress, held in 1982, family planning was included as one of the basic state policies in the Constitution of the People's Republic of China, which reads, 'The state promotes family planning so that population growth may fit the plans for economic and social development.' (Policy Department of the State Family Planning Commission, 1992).

National population programme

The family-planning programme is the core of China's national population programme and is aimed at regulating the number and spacing of births, a factor directly related to population growth. Several other population-related issues, such as marriage, migration, urbanization, population census and data collection, are also addressed in the national population programme. To co-ordinate other agencies and non-governmental organizations (NGOs), such as the Ministry of Public Health, the China Family-Planning Association and the All-China Women's Federation, the State Family-Planning Commission (SFPC) was established in 1981 and is the principal governmental agency responsible for the implementation of China's population programme.

There are many special programmes implemented by various government agencies and NGOs that support the national population programme. These programmes address various areas related to population and development, such as poverty alleviation, maternal and child health (MCH), human resource development, old-age support, environmental protection etc. With the integration of the population programme into socio-economic planning, these programmes will play more and more important roles in promoting China's national population programme.

Modernization and population targets

After the Cultural Revolution (1966–76), which almost led the national economy to collapse, the government took the development of the country's economy as its first priority and started a modernization drive. A three-step strategic plan was worked out under the leadership of Deng Xiaoping. The first step was to double China's gross national product (GNP) in the 10 years between 1980 and 1990. The second step was to quadruple China's per capita GNP by the year 2000. Under this plan it was projected that living standards should significantly improve even if the national population increased by 300 million between 1980 and 2000. The third step involves China doubling its GNP from the 2000 level by 2010 (*People's Daily*, 1996). It is obvious that these strategic targets were based on two principles, one being to develop the economy, the other, to slow down population growth. The 'three-step' strategy sets the targets for both national economic planning and population planning. As a result of it, the size of the national population should still be below 1.3 billion in 2000, and below 1.4 billion in 2010.

In the late 1970s people in China knew little about demographic transition and the laws of fertility decline. Faced with economic difficulties, more attention was given to curbing population growth than to increasing per capita income. Some experts suggested that the total population should be controlled below 1.2 billion and the population growth rate should reach zero by the year 2000. Soon the 'One Child Per Couple' policy was proposed to reach this population target. The government adopted the one-child policy in the late 1970s, but met with strong resistance and great difficulties. In fact, the advocated low-fertility target was not realized as expected. The unrealistic population target (1.2 billion by the year 2000) was abandoned after 1984. A revision of the population target was approved in the late 1980s as part of the Eighth Five-Year Plan (1991–5) for National Economic and Social Development.

The current national population plan approved by the National People's Congress in 1991 is to keep the average natural growth rate of population at about 12.5 per thousand during the 1991–2000 period, and the total population in China below 1.3 billion by 2000. It requires a reduction in the total fertility rate from 2.3 in 1990 to 2.1 in 1995, and to below 2.0 at the end of the twentieth century (Peng, 1992).

Fertility policy

The current family-planning policy can be described in the following terms: to promote late marriage and deferred childbearing, to encourage people to have fewer but healthier births, to promote the practice of 'one child per couple' and to encourage a longer space between births for couples who have practical difficulties if they only have one child. Actually, the current family-planning policy took a considerable time to formulate and develop, and the intention to make further improvements to it has never been given up, even though the aim of 'keeping family planning policy stable' is repeated each year by the government (CPIRC, 1995).

In the period 1949–53 China adopted a pronatalist policy copied from the Soviet Union, granting allowances to couples with many children, prohibiting induced abortion and sterilization and banning the introduction and production of contraceptives. During the period 1954–66 the attitude of the government to contraception changed, to a certain extent, and abortions were allowed under certain conditions. In 1962, with the ending of the 'three-year economic disaster' (1959–61), the extremely high fertility

that compensated for the extremely low fertility of the disaster years caused
the government to promote family planning in highly populated urban
areas. However, when the Cultural Revolution started in 1966, the family-
planning effort was interrupted. As a result, some of the family-planning
organizations were shut down.

The current family-planning programme and its policy started in the
early 1970s. The national policy of delaying marriage and childbearing,
spacing births, and having fewer children (*Wan-Xi-Shao*) was implemented
in 1973. The lowest age for marriage was set at 25 for males and 23 for
females; a two-child norm was promoted; and a birth interval of at least
four years was highly recommended. But this policy was soon replaced by
the one-child policy in 1979. The one-child policy met resistance from
couples with only one child, especially if the first child was a girl, so amend-
ments were soon made. The policy of 'opening a small hole' was adopted
in 1984. Such a policy allows couples to have a second child under certain
conditions. For example, it allows rural couples in some areas with only a
girl to have a second birth after an interval of several years. Since then, the
stabilization of the current family-planning policy has been advocated.

Policy variations among regions

To accommodate the gap between national or collective interests and the
preference of individuals and the significant variations in social, economic
and cultural conditions across regions and among different groups of
people, the local authorities have been given some flexibility in adapting the national policy to local conditions when formulating local family-
planning policies and regulations.

Local fertility policies can be grouped into four major categories:

1. A one-child policy with very few exceptions allowing couples to have
    two children was practised in the three municipalities of Beijing,
    Tianjin and Shanghai, as well as in Jiangsu province and part of
    Sichuan province. The total population in these municipalities and
    provinces makes up about 17 per cent of the national total. This
    policy is also applicable to all residents of urban areas across China.
2. A 'two children if the first one is a girl' policy was implemented in
    rural areas in 18 provinces or autonomous regions accounting for 70
    per cent of China's total population.
3. A 'two children with a four-year spacing' policy was adopted among
    the rural population in the provinces of Guangdong, Hainan and
    Yunnan, as well as in parts of Hebei and Hunan provinces. It is esti-
    mated that the population living in these areas represents about 10
    per cent of the national total.
4. A two-or-three-child policy was adopted in autonomous regions
    inhabited by minority populations, such as Tibet, Inner Mongolia,

Ningxia and Xinjiang, and some small autonomous prefectures in other provinces. It allows the minority ethnic populations to have two or three children. In Tibet, for example, contraceptive services are available only to rural Tibetans who already have three or more children.

By January 1991 there were 28 municipalities, provinces and autonomous regions where local family-planning policies had been formulated and promulgated by local People's Congresses, the only exceptions being Xinjiang and Tibet which adopted their own specific fertility regulations worked out by local government. Targets, principles and strategies for family-planning programmes at all levels have been set according to the family planning policies.

The Family-Planning Programme

The SFPC takes responsibility for the design and implementation of the National Family-Planning Programme, and for supervising 30 provincial family-planning committees. (Tibet has a family-planning office affiliated to the Bureau of Public Health that takes charge of family-planning work in the autonomous region.) The provincial committees and all their subordinate agencies at the county, township and village levels form a well-knit family-planning network throughout the country (see table 5.1).

The responsibility system

In the early 1990s, the Central Committee of the Chinese Communist Party and the State Council introduced the family-planning responsibility system. The system requires that heads of Party organizations and governments in all provinces, autonomous regions and municipalities take full responsibility for implementing their local population plans, integrating population plans with their social and economic development plans, and giving priority to the family-planning programme. Government at all levels is to implement and improve the responsibility system of population-target management, and leaders of Party committees and governments at all levels are personally responsible for the accomplishment of their population plans. Failure to meet the population-control targets may be subject to some kind of penalty, such as the withholding of a bonus, demotion or dismissal.

The yearly plans for the family-planning programmes at all levels should be worked out according to the national family-planning programme, which was developed according to the national five-year population plan and long-term target. These programmes are evaluated and assessed each year primarily on the basis of quantitative measurements such as the number of births, the crude birth rate, the contraceptive prevalence rate (proportion

Table 5.1 Family-planning organizations in China

Admin. level	Executive branch	Party unit	Health unit	Family-planning unit
Nation	State Council	Politburo	Ministry of Public Health	State Family-Planning Commission
Province, City or Autonomous Region	Provincial Government	Provincial Party Committee	Provincial Health Dept; Anti-epidemic & MCH Station, Hospitals	Provincial FP Commission
County	County Government	County Party Committee	County Health Dept; Hospitals; Anti-epidemic & MCH Station	County FP Commission; FP Service Station
Town or Township	Town/Township Government	Township Party Committee	Town Health Central	Township FP Committee; FP clinic*
Village	Village Leadership Group	Village Party Branch	Co-operative Medical Station	FP Leading Group; FP post*
Group or Team	Group Leader	Party Members	Part-time Health Aide	Part-time FP Worker

*The FP clinics or posts at township and village have not been completed yet.

of couples using contraception among all couples of reproductive age) and the rate of planned births etc.

There is little doubt that the responsibility system of population-target management has strengthened leadership in implementing the policy, mobilizing resources and co-ordinating activities for the family-planning programme. The input by governments at all levels for family planning increased through the early 1990s, from 1.5 billion yuan in 1991 to 2.7 billion in 1995 (CFPYB Editorial Office, 1992; CFPYB Editorial Office, 1995). It is estimated that the per capita spending on family planning exceeded 2.64 yuan in 1995 (about one-third of one US dollar) and will reach 4 yuan by 2000 (Peng, 1996). However, the strong emphasis on meeting the birth quota has also resulted in some side effects, such as coercion and statistical falsification (exaggerated achievements and under-reported births).

In recognizing the limitations of the target-oriented programme, the government began to emphasize: (1) information, education and communication (IEC) instead of administrative pressure; (2) contraception instead of induced abortion; (3) regular family-planning services instead of family campaigns. Such an approach was first adopted in some provinces in the early 1980s. In mid-1993 the SFPC commended and gave awards to 100 model counties for implementing the service-oriented family planning programme in order to encourage others to follow their example.

Contraceptive services

Contraceptive services in China are mainly aimed at married couples, especially those who already have a child. Most married couples at reproductive age receive contraceptive services free of charge. Different contraceptive methods are recommended to couples according to the number of children that they already have. The intrauterine device (IUD), for example, is suggested to couples with only one child, and sterilization is recommended to those who have had two or more children and do not want any more births. Following Chinese tradition, newly married couples usually have their first baby within one or two years of marriage, and very few use contraceptives. They are not, therefore, regarded as the major group within the target population. Unmarried people are usually excluded from the target population for contraceptive services. However, the increasing prevalence of premarital sex and abortions is making it necessary to expand family-planning and reproductive services to the unmarried and newly married.

During the 1970s the contraceptive prevalence rate (CPR) was no more than 60 per cent. It increased to around 70 per cent in the early 1980s and stood around that level in the 1980s, increasing only very slightly from 69.5 per cent in 1982 to 71.2 per cent in 1988. In the early 1990s the CPR increased further, reaching 83.4 per cent in 1992, much higher than the average levels in both developed countries (72 per cent) and developing countries (53 per cent including China) (United Nations, 1994). However, there are regional variations and still unmet needs in remote areas, especially in mountain areas (see figure 5.1).

Most couples in China use highly efficient long-term methods, such as the IUD and sterilization, which are strongly endorsed by government policies and regulations. According to a United Nations estimate, the contraceptive method mix in 1992 among Chinese users was 41 per cent female sterilization, 12 per cent male sterilization and 40 per cent IUD. That is, sterilization and IUD constitute 93 per cent of all methods being used, and less than 7 per cent of the contraceptive users use the pill, condoms and other modern or traditional methods. This contraceptive-use pattern is very different from that of the developed countries where only 25 per cent of acceptors use provider-controlled methods, 22 per cent use the pill, 19 per cent use condoms, and more than 30 per cent use traditional methods (United Nations, 1994). Both government family-planning policy and regulations and the difficulty of maintaining a continuous supply of short-term methods and ensuring user compliance contribute to such a high prevalence of sterilization and IUD use. The increasing CPR has led to a steady decline in

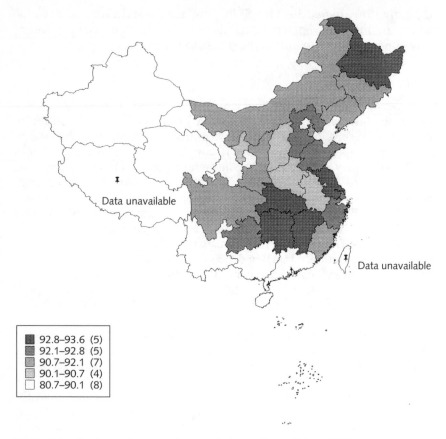

92.8–93.6 (5)
92.1–92.8 (5)
90.7–92.1 (7)
90.1–90.7 (4)
80.7–90.1 (8)

Figure 5.1   Contraceptive prevalence rate by regions, China, 1995

abortions from about 20 million per year in the 1980s to 10 million in 1992. The reported abortion ratio is less than 0.5 (CFPYB Editorial Office, 1993).

Information, education and communication (IEC) services

For more than 20 years, population education has been given both in and out of school and has contributed to the dissemination of information and knowledge about family planning. Repeated publicity about the family-planning policy and regulations and introducing contraceptive knowledge has brought family-planning issues into the public arena. Deciding that 'it's good to practice family planning' has become a new and prevailing custom, and the advocacy of a 'cultured and happy small family' has become a fashion. Family-planning workshops are often conducted by the local family-planning commissions, neighbourhood committees, the youth league and the women's federation, free of charge.

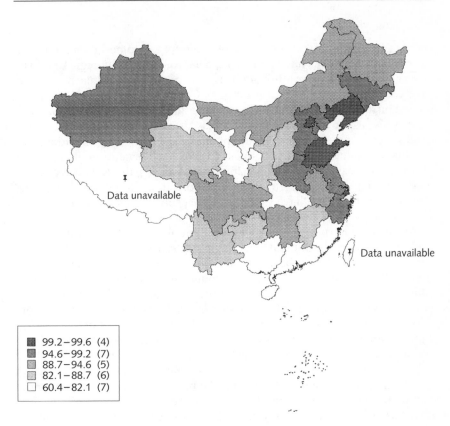

Data unavailable

Data unavailable

■ 99.2−99.6 (4)
■ 94.6−99.2 (7)
▨ 88.7−94.6 (5)
▨ 82.1−88.7 (6)
□ 60.4−82.1 (7)

Figure 5.2   Rate of planned births by regions, China, 1995

Mass media, mostly in the form of television, newspapers and radio broadcasting, often display urban lifestyles, where the one-child policy has been in effect for more than 18 years since 1979. Television programmes from Hong Kong, Taiwan and western society not only increase the desire for material wealth, but also introduce the value of small families. All these influences contribute to changing fertility desires (Wang and Ola, 1994).

The IEC programme of family planning in China used to rely heavily on the mass media and public workshops, focusing mainly on the dissemination of information about government family-planning policy and regulations and the justification of a strong family-planning effort, while individual face-to-face counselling about contraceptive knowledge and reproductive health issues was relatively neglected. However, in recent years there has been an increasing awareness of the importance of personal counselling and more and more family-planning clinics have begun to provide individual counselling.

Incentives and disincentives

As measures to enforce the family-planning programme in China, incentives and disincentives have played a special role in fertility regulation. The government grants a monthly allowance to single-child-certificate holders (five yuan for a boy, six yuan for a girl). Couples promising not to have a second birth after having one child can get a single-child certificate and they have access to the allowance from the day the certificate is issued until the child reaches age 14. Local family-planning agencies also encourage the parents' work units or village councils to give some benefits or gifts (money or articles for daily use) to single children and their parents who accept the certificate. Up to the present there are more than 50 million single-child families in China that have accepted the certificate and received various kinds of allowances and benefits.

Arable land in rural China is the most important natural resource, especially in the densely populated areas in eastern China with a per capita share of cultivated land below 0.3 *mu* (about 2 ares). Since the introduction of the household production responsibility system in 1980 in rural China, each farming household gets a piece of contracted land whose size is based on the number of its members. Therefore, a family with more members can get a larger share of farming land. This kind of practice works against family planning. In order to encourage families to have fewer children, some local authorities give single-child families a double share of arable land and land for building houses. Families having an extra birth may be denied the share of land. With the booming market economy, more and more young rural people have left farming to seek jobs or run their own business in non-agricultural sectors. Disincentive measures, such as allocating less cultivated land, consequently have less effect on the fertility behaviour of farmers.

The disincentives to urban residents or to people with a permanent job have proved more effective. Violating the family-planning policy or regulations not only means facing the danger of losing such benefits as the right to acquire a flat and subsidies for children's education, but also poses problems with employment and promotion.

Opportunities and challenges

The success of China's population policy and programme is well documented. Nevertheless, there is significant regional variation. All the above-mentioned factors – policy, regulations, service, government commitment – have had an influence on the achievement of birth control targets in local areas. The regional picture can be seen in figure 5.2, where the rate of planned birth is used as the indicator. Generally speaking, the family-planning programme has achieved more success in the eastern coastal regions and been less successful in the inland provinces.

With the ongoing economic reform characterized by decentralization and increasing individual mobility and choice, the family planning programme is facing new challenges and difficulties. Some rich farmers are not afraid of paying off the fine for extra births. The increasing numbers of men and women migrating from rural to urban areas also pose great difficulties for family-planning officials and workers who have to provide family-planning services and enforce the family-planning regulations. However, economic development has also provided new opportunities for family planning. Increased incomes and job opportunities have also raised the opportunity costs of having children. More and more men and women at reproductive age, especially around their twenties, are involved in non-agricultural sectors, and the time spent on childrearing means lost money and career opportunities. Economic reform in China has enhanced the living standards of farmers and, as a result, changed their values and fertility attitudes, especially those of the younger generation.

Meanwhile, the transfer of labourers from rural to urban areas and from agricultural to non-agricultural sectors is having a significant impact on fertility. Having young children or being pregnant are seen as incompatible with the search for better jobs and opportunities for migrants. That leads to postponed marriage and childbearing. Spending on education, health care and daily life for children have increased dramatically with the improvement of living standards. This means that some young parents want fewer children. The expense of marriage is also a major determinant in the timing of marriage and childbearing. The cost of marriage in the 1990s, for example, in eastern China where the economy has boomed, ranges from 50,000 to 100,000 yuan according to a recent survey, almost 10 times the cost in the 1980s. This means many years of saving by the young couple and their families are necessary before they can afford the marriage.

A New Approach in Family Planning

Faced with the changes in socio-economic development and the control of population growth, the SFPC proposed a new approach in 1995, called the Integrated Approach to Family Planning with Development, as one of the government actions following the United Nations International Conference on Population and Development in Cairo in 1994. The expanded population programme not only emphasizes control of population quantity, but also pays attention to the improvement of the quality of life of the population and control of its structure, including its age structure, sex composition and geographic distribution. The fundamental aim is to create a favourable population environment for modernization and the improvement of people's living standards.

In shifting from a demographic orientation to a service orientation, the family-planning programme in rural China should be integrated with the developing rural economy, helping farmers to become prosperous and building 'cultured and happy families'. Women-centred quality services for production, daily life and childbearing (*Shengchan, Shenghuo, Shengyu* in Chinese) will be delivered to family-planning acceptors and poor families to help them have fewer births and create wealth.

The Chinese government has stressed almost every year since 1988 that the population policy should not be changed, that the population target must remain in place, and that leaders must continue to take responsibility for family-planning implementation, in order to show its determination and its commitment to implementing the population programme. In fact, however, the policy itself, its targets and the approaches to practising population and family-planning programmes have undergone many changes or improvements in the past two decades.

While successful policy implementation depends largely on the acceptability of the policy itself, the maintenance of a large staff of family-planning officials and workers in China has remained essential. There are about 300,000 officials and workers working in the family-planning system, of whom more than 200,000 work at the level of towns/townships or below. There are also hundreds of thousands of part-time family-planning workers in villages. With a family-planning network established, married couples at reproductive age have easy access to family-planning services. Meanwhile, there are at present 900,000 branch associations under the China Family-Planning Association throughout the country with more than 50 million members assisting the implementation of the family-planning programme (China Family-Planning Yearbook Editorial Office, 1993).

References

China Family Planning Yearbook (CFPYB) Editorial Office (1993), *China Family Planning Yearbook* (in Chinese). Beijing: CFPYB Editorial Office.

China Family Planning Yearbook (CFPYB) Editorial Office (1995), *China Family Planning Yearbook* (in Chinese). Beijing: CFPYB Editorial Office.

Chen, Shingle and Shao, Wei (1995), *A Serial Report of the Family and Contraceptive Sample Survey Chart Volume.* Beijing: China Population Publishing House.

CPIRC (1995), China's family planning policy and population development. In *Monograph on the 1990 Population Census of the People's Republic of China* (Vol. 14), edited by SSB, DPE. Beijing: China Statistics Press.

Gu, Baochang and Peng, Xizhe (1991), Consequences of fertility decline: Cultural, social and economic implications in China. Paper presented at KIHASA/ESCAP Seminar on Impact of Fertility Decline on Population Policies and Programme Strategies: Emerging Trends for 21st Century, 16–19 December 1991, Seoul.

Peng, Peiyun (1992), The Population of China: Problems and Strategy. *China Population Today*, 1994 (4).

Peng, Peiyun (1995), Population and Development in China. Speech to the Science Conference of the Asia-Pacific Region, Beijing.

Peng, Peiyun (1996), Speech to the family-planning forum held by the Central Committee of the Chinese Communist Party on 10 March 1996.

People's Daily (1996), The People's Republic of China National Programme of the Ninth Five-Year Plan and 2010 Long-Term Target for Economic and Social Development. *People's Daily*, 20 March.

SFPC, Policy Department (1992), *Family-Planning Documents Collection 1981–1991*. Beijing: China Democracy and Legality Press.

United Nations (1994), *World Contraceptive Use, 1994*. Population Division, UN.

Wang, Feng and Ola, Nygren (1994), Miracles and myths: China's below replacement fertility (incomplete draft).

Xie, Zhenming (1994), Regarding men as superior to women: Impacts of Confucianism on Family Norms in China. *China Population Today*, 1994 (6), 12–16.

Xie, Zhenming (1995), Report on Tibetan Population Development and Family Planning. *Population Research*, 1996 (1), 41–8.

Yao, Xinwu and Yin, Hua (1994), *Basic Data of China's Population*. Beijing: China Population Publishing House.

6   Age and Sex Structures

Li Yongping and Peng Xizhe

The age and sex structures of a population reflect the history of its natural demographic process of birth, death and migration, as well as societal influences such as wars and famines. Also, age and sex structures imply the future development of a population, providing important information for governments to use in planning various population-related policies. Mainly using the data of the 1990 census, this chapter will study the age and sex structures of China.

Age Structure

There are three main factors that have caused changes in the age structure in China since the founding of the People's Republic in 1949. The first is the rapid decline in the death rate, particularly the infant and child mortality rates, because of a dramatic improvement in the health care and hygiene of the masses. It has resulted in there being greater proportions of both elderly people and young children in the population. The second factor is the huge impact of socio-economic changes. Socio-economic changes have brought about migrations of different magnitudes and in different directions, for example, from rural to urban areas and from inland to coastal areas, especially after the economic reform started in 1978. The rapid improvement in living standards after 1949 also led to a sharp population increase. The third factor is the historically unprecedented rapid fertility decline, due to the stringent family-planning policy. Family planning has been leading to the gradual shrinkage of the base of China's population pyramid.

China's high quality of age reporting for age-specific data is due to the fact that almost all Chinese know the 12 animal years and which they were born in. The data quality of the 1990 census is reasonably good and consistent (Qiao, 1992). Based on the data for population distribution by age and sex in single-year age groups, figure 6.1 illustrates the 1990 population age pyramid.

On 1 July 1990 there were 1.13 billion people living in China, of whom 0.58 billion were females and 0.55 billion males. From figure 6.1, it can be

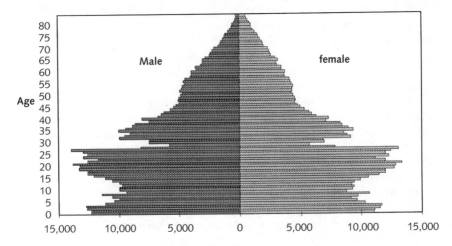

Figure 6.1 China population pyramid, 1990

seen that there are several irregularities in the age structure. The first irregularity is the narrow but deep indent in the age group 28–31. The relatively small number of people in this age group was the result of the famine of 1959–61. During those three harsh years when people faced starvation, there were fewer births (Peng, 1987).

The second irregularity is the wide deep indent for the age group 4–16. This relates closely to the change in population policy. China began its effective campaign of family planning, '*Wan* (later marriage), *Xi* (longer birth interval), and *Shao* (fewer births)' in the 1970s. Most of the indent above age 8 is the result of the reduction in the number of births during the period of this campaign. There was a temporary increase in the number of births in 1982, as it can be seen in age group 8. This was a period effect caused by the relaxation of the marriage law in 1982 that led to a temporary reduction in age at marriage and childbirth. Period fertility increased, but cohort fertility did not. In the 1980s China further strengthened its population policy to cope with the dramatic increase in the number of people of reproductive age. This new policy was known as the 'one child per couple' policy (for a detailed discussion see chapter 5). Most of the indent under age 8 can be attributed to the implementation of the one-child policy. Looking at the age pyramid, one might conclude, erroneously, that the *Wan, Xi, Shao* policy was just as effective as one-child policy, if not even more so. It should be borne in mind, however, that in the period of *Wan, Xi, Shao* there was a smaller cohort of young reproductive women than during the period of the one-child policy. In 1984 the Chinese government relaxed the regulations on family planning and modified the one child policy. The main distinguishing feature of the 'relaxed one-child policy' is that a family in a rural area can have a second birth if the first birth is a female. The impact of such a policy relaxation can be vividly seen from the rapid increase of population in the age group 0–4 in Figure 6.1. However, the population policy is more stringent in urban areas where the favourable attitude towards large families is not so prevalent because of the influence of modern societal values.

Between these two indents there is a bigger cohort in the age group 16–27, whose development in the future is crucial. For example, because of its relatively large size, the strictness of family-planning policy can hardly be relaxed within the reproductive age span of this cohort. It is unlikely, therefore, that the 'relaxed one-child policy' will be further relaxed before 2010, if the government wishes to protect the environment and prevent the population from exploding. The size of this cohort has determined the nature and reality of Chinese family planning. In practical terms this means a bigger supply of contraceptives and a greater demand for prenatal services.

China's total fertility rate in the 1990s has been very low, about 2 children per woman, which is below the replacement level. However, because there is a large female population of reproductive age, China's population is still increasing. This phenomenon is called the influence of the existing age structure, or population momentum.

The problem of population ageing is emerging. Owing to the large relative as well as absolute size of the 16–27 cohort, as it develops into an elderly population around 2030, China will enter an unprecedented stage of population ageing. The social security system will have to bear a heavy burden and face great challenges. While most of the urban population benefit from a pension system, most of the rural population have to resort to the traditional method of obtaining support and security during old age, looking to the family, especially a son. The Chinese government must plan a way to transfer successfully from the current 'pay-as-you-go' and family support systems to a sound modern system of pensions plus savings.

Arranging the population age groups into child dependency (0–14), labour force (15–64), and elderly dependency (65+), table 6.1 compares China's dependency with that of the rest of the world. In 1990 China's child dependency and elderly dependency were both lower than those of other countries. This means that China had a relatively large labour force. However, China had a lower percentage of aged population (46.7 per cent lower) and a higher percentage of child population (25.9 per cent higher) than the developed countries. This shows China's age structure of population is still a young one, even though fertility is below the replacement level.

Longitudinally, table 6.2 shows the development of the dependent sections of China's population from the first census in 1953 to the fourth census in 1990. After the foundation of the People's Republic of China in 1949, China's population of dependent children experienced first two stages of increase and then a decrease, the decrease in recent years being especially

Table 6.1 Comparison of population dependencies (%), China and world, 1990

Region\Age	0–14	15–64	65+
China	27	67	6
World	33	61	6
Developed countries	22	66	12
Developing countries	36	60	4
England	19	66	15
USA	22	66	12
India	39	58	3
South Korea	27	68	5

Source: World Population Data Table (PRB, 1991).

Table 6.2 Comparison of population dependencies in four census years, China

Age\Year	1953	1964	1982	1990
Age structure				
0–14	36.3	40.7	33.6	27.7
15–64	57.3	55.7	61.5	66.7
65+	4.4	3.6	4.9	5.6
Dependence ratio				
0–14/15–64	0.63	0.73	0.55	0.42
65+/15–64	0.07	0.06	0.08	0.08

Sources: China Population Year Book (PRICSA 1985); 10 per cent Sampling Tabulation of the 1990 Population; Census of the People's Republic of China (CSB 1991).

notable. Since 1964, while dependent-child population has been decreasing, both the adult population and the elderly population have been continuously increasing. This implies that China's population is currently transferring rapidly from a young age structure to an adult age structure, and then to an old age structure. Compared with the relatively slow demographic process of developed countries, this trend in China seems very fast and irreversible.

The child dependence ratio has been decreasing quickly – from 0.73 in 1964 to 0.42 in 1990. That is, one adult needed to support 0.73 children in 1964, but only 0.42 in 1990, a relative decrease of 42 per cent. This is the most obvious success of family planning. On the contrary, as a kind of compensation, elderly dependency has been increasing slowly from 0.06 in 1964 to 0.08 in 1990. In terms of a qualitative assessment, child dependency implies mainly the input of education, elderly dependency the input of health care. Education and health care are two major governmental expenditures whose budget appropriations are closely related to the changing age structure.

Family-planning policy exerts a strong influence on the aggregate age structure. The age structure of China's population is also, however, greatly affected by variations between rural and urban areas, variations between regions of different economic and cultural development, by the presence of ethnic minorities, and so on. These factors are responsible for its complexity and multiplicity. With respect, for instance, to the relation of age structure to geographic locality, the transition of population from high to low birth and growth rates is quicker in the south and east along the coast than in the southwest and northwest (Wu, 1984). To take another example. Owing to migration and a preferential economic policy, the age pyramid in Shenzhen city (a close neighbour of Hong Kong) shows a 20 per cent excess of young women aged 15–24 in 1990. As might be expected, this abnormal age structure, extremely favourable for young men, is having a strong influence on the local marriage market, resulting in many social and economic problems (Guo, 1996).

Sex Structure and Sex Ratio at Birth

In 1990 the sex ratio of the total population was 106 males to every 100 females. Figure 6.2 depicts the sex ratios of single-year age groups in 1990. Numbers reveal that child age groups (below age 5) had high sex ratios, adult age groups (between ages 40–60) high sex ratios, and elderly age groups (above age 70) low sex ratios. The low sex ratios for the old age group, such as 85.8 males per 100 females for age group 70–4, is due to differential mortality. In almost every country including China, females in old

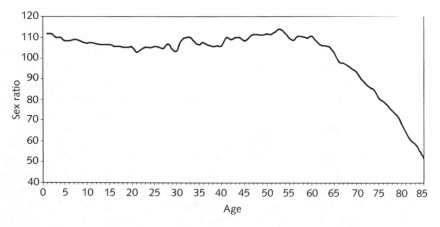

Figure 6.2   Age-specific sex ratio of China, 1990

age have a higher life expectancy than males of the same age. The high sex ratio of the adult age group is most likely a result of the Japanese invasion (lasting 8 years) and the civil war (lasting 3 years), during which periods females suffered a much higher death rate. The high sex ratios for young age groups such as 110.2 males per 100 females for the age group 0–4 are, however, unusual. This abnormal ratio is the focus of the so-called 'missing girls issue'. If these girls are really missing what are the direct or indirect causes? The issue of missing girls is at root an issue of missing female births. The highly abnormal sex ratio at birth in China is of major concern to the balance of the sex structure, as well as pointing to possible discrimination against females at birth.

Figure 6.3 presents a time series of sex ratio at birth from the 1950s up to the early 1990s. It shows that the reported sex ratio at birth in each year of the 1980s was abnormally high, and that the sex ratio at birth apparently increases with the total number of children born (Hull, 1990). In addition, research reveals that the abnormal sex ratio is not an entirely new phenomenon emerging after the 1980s, rather that it is a continuing one that happens to be more visible at the present time. Deliberate intervention to control the sex of children has existed through the past several decades, at least in certain groups (such as groups with no son or few sons). The number and the sex structure of existing children are the two major factors affecting the sex ratio, especially at high parities.

The abnormality of sex ratio at birth (SRAB), however, is not a phenomenon unique to China. The nature of the problem is likely to be the same in many developing countries, but the magnitude and consequences may be different. There is great similarity in terms of patterns of SRAB between China and South Korea. Generally speaking, the SRAB increases

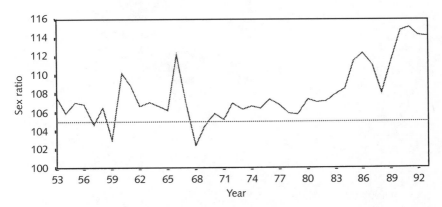

Figure 6.3   Sex ratio at birth in China, 1953–1993

Sources: Data for 1953–88 calculated from two per thousand National Fertility and Contraception Survey, 1988, by State Family planning commission; 1989 from 1990 population census; 1990–3 from Annual Sample Survey for Population Change by State Statistical Bureau of China.

by birth order in both countries. While the ratio for first children can be regarded as normal, it becomes very high for third and higher-parity births and has increased over the 1980s. Moreover, the sex ratio for higher-order births in South Korea is even higher than that in China (Chai Bin Park and Nam-Hoon Cho, 1995). A similar situation has been found in Taiwan, and is also seen in Hong Kong, India, Pakistan, Bangladesh and the region of west Asia (see, for example, Das Gupta, 1987; ESCAP, 1990).

The abnormality in sex ratio at birth is a nationwide phenomenon in China in the 1980s. Nevertheless, there are tremendous differences among the different sub-regions of mainland China. Among 30 provincial administrative units, according to China's 1990 population census, high sex ratios prevailed in 24 provinces, and only 6 reported figures more or less within the normal range (table 6.3 and figure 6.4). At the lower administrative levels, based on an incomplete data source, 232 out of a total of 334

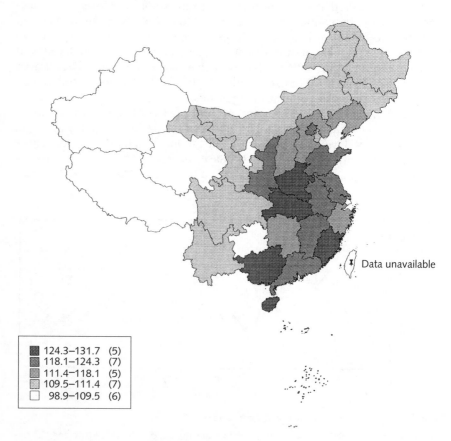

■ 124.3–131.7	(5)
▦ 118.1–124.3	(7)
▨ 111.4–118.1	(5)
▧ 109.5–111.4	(7)
□ 98.9–109.5	(6)

Figure 6.4 Sex ratio for age zero, China, 1995

Table 6.3 Sex ratio at birth and for other age groups
in 1995 and 1990 by provinces

	Sex ratio at birth		Sex ratio for other ages in 1995	
	in 1995*	in 1990	1–4	5–9
China	116.57	111.42	118.81	110.19
Beijing	122.54	106.21	111.62	105.94
Tianjin	110.56	110.65	109.91	107.22
Hebei	115.20	112.32	115.66	106.34
Shanxi	111.83	109.66	112.42	109.13
Inner Mongolia	111.36	107.37	113.17	106.96
Liaoning	111.61	110.10	114.97	108.07
Jilin	109.84	108.11	109.92	108.71
Heilongjiang	109.70	107.44	107.20	105.96
Shanghai	105.34	104.35	102.60	102.49
Jiangsu	123.88	114.50	123.27	111.40
Zhejiang	115.35	117.82	115.19	110.74
Anhui	118.14	110.48	126.71	113.11
Fujian	124.42	110.49	127.97	109.62
Jiangxi	119.81	110.56	126.32	109.68
Shandong	118.94	115.97	120.72	110.97
Henan	127.44	116.64	133.14	111.80
Hubei	131.63	109.49	123.29	109.94
Hunan	116.96	110.49	118.97	108.21
Guangdong	123.30	111.76	119.76	113.83
Guangxi	124.57	117.73	128.71	118.81
Hainan	125.87	115.60	121.52	116.17
Sichuan	110.01	111.53	111.79	109.50
Guizhou	100.35	101.77	115.96	111.33
Yunnan	109.53	106.84	110.78	108.93
Xizang	98.91	103.05	105.29	101.91
Shaanxi	124.26	111.12	118.56	111.63
Gansu	110.13	110.29	116.40	107.32
Qinghai	106.58	104.62	105.53	105.85
Ningxia	106.77	110.04	108.55	105.49
Xinjiang	101.26	103.70	101.91	103.82

*Sex ratio at birth here refers to the ratio for zero age group.
Source: Figures for 1995 are calculated from *Data of 1995 National 1% Population Sample Survey*, pp. 2–10, Beijing: China Statistical Publishing House, 1997; 1990 data are derived from Population Census Bureau, *10% Sample Tabulation of China's 1990 Population Census*, Beijing: China Statistical Publishing House, 1991.

prefecture units (70 per cent) and 1,301 out of 2,067 counties (63 per cent) witnessed higher than normal SRABs.[1] Geographically speaking, the high SRABs were concentrated in the southeastern part of China, with the highest, 117, recorded in Zhejiang province. Except for the big metropolitan area of Shanghai, the remaining 5 provincial units where the SRAB was below 107 in 1990 were all located in the western part of China.

In 1990 the population of cities, towns and rural counties accounted for 18.69 per cent, 7.54 per cent and 73.77 per cent of total population in China, respectively. The reported SRAB in 1989 was 108.90 in cities, 112.31 in townships and 111.60 in rural areas. The highest SRAB was found in China's towns rather than in the countryside. At the provincial level, there were 12 provinces whose townships reported the highest SRAB. Given the rapid expansion of towns in the last decade, many town residents are new settlers from neighbouring villages. On the whole, therefore, the abnormal sex ratio at birth is most serious in rural areas.

Causes of Abnormal Sex Ratio at Birth

In recent years, many articles have intensively studied this issue of missing girls, some quite controversially. There are basically three hypotheses to explain the direct causes of this abnormal sex ratio at birth. The first hypothesis is prenatal sex determination by use of an ultra-sound-B machine. The principle use of ultra-sound-B technology is in IUD check-ups. Currently ultra-sound-B machines are available in every county, and even in villages and townships. Because of the existing strong preference for sons in China in consideration of the family lineage and security in old age, people may selectively abort girls in order to have another chance of pregnancy. On this hypothesis, both supply and demand operate, leading to an actual surplus of boys. The second hypothesis is the under reporting of female births (Tu and Liang, 1993) and the adoption of female babies (Johanssen and Nygren, 1991). This hypothesis assumes the 'missing' female births still occur but do not appear in the data. Under reporting and adoption may have played a major role in the past, as some studies showed, yet continuous prevalence is problematic. The third hypothesis is the higher mortality of female infants and children (Coale, 1991). That is, female infanticide may exist as a traditional practice in some areas. Additionally, lack of care and mistreatment of female babies may also result in a high female infant mortality. Unfortunately, this hypothesis has focused only on the statistical figures and ignored the practical implications of the high physiological cost of a full-term pregnancy. Assessing the costs and benefits, together with the study of time series data, suggests the first hypothesis of selective abortion is more plausible, although these hypotheses are not necessarily mutually exclusive.

Abnormal sex ratio at birth is a complicated phenomenon resulting from overall socio-economic settings. A reported SRAB that is abnormally high results both from a real imbalance due to the disturbance of social factors and from the wide practice of underreporting female births. The real magnitude of the abnormality has been exaggerated by statistical error. Nevertheless, among many possible causes, the stages of the fertility transition and the cultural tradition of son preference may be among the most relevant to the phenomenon, which needs to be discussed in some detail.

The emergence of prenatal sex selection is mainly due to the imbalance between people's desire for a particular number of children and their desire for children of a particular sex. In other words, the abnormally high SRAB results directly from the inconsistency of social, economic and cultural development, with the decline of fertility and the improvement in modern medical and contraceptive technology. Whether fertility declines due to the implementation of birth control policy by the government, or people's voluntary self-control of fertility does not make a radical difference, as can be seen from the typical examples of China and South Korea. The nature of the abnormality problem in many developing countries is likely to be the same, but its magnitude and potential consequences are different.

The transition from a centrally planned economy to a socialist market economy, and the consequent changes in social and economic environment have to some extent made Chinese women more vulnerable to discrimination, which again strengthens son preference. At the same time, the family-planning programme has facilitated changes in people's fertility strategy, putting more emphasis on the quality and gender of children in a low-fertility regime. This too has resulted in a stronger manifestation of a son preference.

In a society without a well-functioning social security system, with a low level of economic development and correspondingly scarce economic opportunities, Chinese peasants have to rely heavily on their family as their basic survival strategy. Thus, in addition to traditional patriarchal ideology, for most families in rural China there is still a big difference between the real-life functions of sons and daughters. Males are the people expected by society to be responsible for the continuation of the family name, the alleviation of all the life risks that the family faces, etc (Gu and Peng, 1992). Consequently, male children often get better opportunities for food, education and medical care. This may be also true in many other developing countries.

As modernization, social policy and other socio-economic constraints reduce the number of children a couple wish to have, the demand for children of higher quality increases. To put it another way, the emphasis in reproductive strategy shifts from quantity to quality in the process of rapid fertility decline (Gu, 1992). In a society with strong son preference, boys are naturally regarded as better than girls in terms of 'quality'. Therefore, after the fertility rate has declined to a quite low level, the preference for sons will seem to be much stronger and more visible, even though people's concepts of sex preference have not changed. In the context of low fertility, people may pursue every available method to realize their son preference, since there are fewer chances for them to fulfil their goals of family formation. This is certainly a dilemma resulting from the fact that changes in the social, cultural and moral system are lagging a considerable way behind rapid technological development. It also reminds us of the persistence of the cultural tradition.

The phenomenon of declined fertility with strengthened son preference is likely to be one of the special stages of fertility transition in societies with strong son preference. Therefore, the fertility transition should not be examined only with respect to numerical indicators; the dimension of sex structure should be seen as an important one needing even more attention.

Summary

China's age structure is experiencing a rapid transition from a very young age structure in the recent past to an old age structure in the near future. Although the current age structure is young, and the population is increasing, the latent threat of rapid population ageing in the next century exists. Both family-planning policy and socio-economic changes have influenced the age structure. The age structures of sub-populations have many different patterns because of different combinations of fertility, mortality and migration.

In 1990, China had a life expectancy figure of about 70 years. In elderly age groups, as expected, there are more surviving females than males. However child age groups tend to have many more males than females. These abnormally high sex ratios at birth and among infants are mainly due to three direct causes: selective abortion of females, underreporting and adoption of female babies, and differential high mortality for female infants and girls. The indirect and underlying cause is the pursuit of high 'quality' of birth under the constraint of low 'quantity' of birth. Though currently there is no problem of imbalance of sex in the marriage market, current age and sex structures imply a severe imbalance in the near future.

With the improvement of monitoring and reporting systems, misreporting of female births has been gradually decreasing since the later 1980s. Moreover, high female infant mortality and incidences of infanticide and abandonment have already drawn attention from both the government and society. People are aware that high and rising SRAB will lead to a serious marriage squeeze in the future which may result in social instability and put more constraints on future economic development and social stabilization.

The government is trying to solve this problem through improving women's status and old age security, eliminating some traditional practices such as lineage and its worship, and strengthening the regulation of ultrasound-B machines. In 'Proposals and Suggestions for Preventing the Rise of Sex Ratio at Birth', a government directive issued in March 1994, the

Chinese State Family Planning Commission says: 'The implementation of the family-planning policy should aim not only at regulating population quantity and improving population quality, but also at achieving a reasonable population structure. If the sex ratio at birth goes far beyond the normal range, the sex structure of the population will be unbalanced, which will result in serious social problems and bring unfavourable influences on population development in the future. In order to take a responsible attitude to our nation and descendants, government officials at all levels are to give serious consideration to this problem and take effective measures so that it can be gradually solved'.

Though the occurrence of reported abnormal sex ratio at birth will decrease, the influence of prenatal sex selection is likely to remain strong for a while. These factors, closely linked to the current abnormal SRAB, are not going to disappear in a short period and the settlement of abnormal SRAB finally will depend on the transition of the social environment as a whole. It will take a long time for most Chinese to replace their feudal consciousness and traditional customs with modern civilized concepts completely in tune with fundamental social and economic development. Nevertheless, it is our firm belief that with the implementation of all those socio-economic measures mentioned above, and with the attention and participation of the whole society, discrimination against women and the abnormality in sex ratio at birth will be gradually eliminated.

Note

1. Calculated from tabulations of China's 1990 Population Census data, higher sex ratio at birth here refers to a ratio above 107. There may be some differences resulting from the classification of urban and rural areas.

References

Chai, Bin Park and Nam-Hoon, Cho (1995), Consequences of son preference in a low-fertility society: Imbalance of sex ratio at birth in Korea. *Population and Development Review*, 21 (1), 59–84.

China State Statistical Bureau (CSSB) (1991), *10% Sampling Tabulation on the 1990 Population Census of the People's Republic of China*. Beijing: China Statistical Publishing House, 93–5.

Coale, Ansley J. (1991), Excess female mortality and the balance of the sexes in the population: An estimate of the number of missing females. *Population and Development Review* 17 (3).

Das Gupta, M. (1987), Selective discrimination against female children in India. *Population and Development Review*, 13 (1), 77–100.

Gu, Baochang (1992), Lun shengyu he shengyuzhuanbian: shuliang, shijian he xingbie (On fertility and fertility transition: number, time and sex). *Renkou Yanjiu (Population Research)*, 6, 1–7.

Gu, Baochang and Peng, Xizhe (1992), Consequences of fertility decline: Culture, social and economic implications in China. In *Impact of Fertility Decline on*

Population Policy and Programme Strategies, Seoul: Korean Institute for Health and Social Affairs.

Guo, Zhigang (1997), Internal migration and population growth of Shenzhen. In *Macau and its Neighbours in Transition*, edited and published by Macau University.

Hull, Terence H. (1993), Recent trends in sex ratio at birth in China. *Population and Development Review*, 16 (1), 63–83.

Johanssen, Sten and Ola, Nygren (1991), The missing girls of China: A new demographic account. *Population and Development Review*, 17 (1).

Li, Yong-Ping (1993), Sex ratio at birth and related social variables: Census results and implications. *China 1990 Population Census – Papers for International Seminar*, edited by China Statistical Bureau, Beijing: China Statistical Publishing House, 249–61.

Peng, Xizhe (1987), Demographic consequences of the Great Leap Forward in China's provinces. *Population and Development Review*, 13 (4), 639–70.

Population Reference Bureau of USA (PRB) (1991), *1990 World Population Data Table*.

Population Research Institute of China Social Academia (PRICSA) (1986), *1985 China Population Year Book*. Publisher of Chinese Social Science.

Qiao, Xiaochun (1992), Initial test of population age and sex structure for the fourth census data. *China Population Science* (in Chinese), 5.

Secretariat of ESCAP (1990), Population situation, policy and programmes in Asia and the Pacific, *Population Research Leads*.

Tu, Ping and Liang, Zhiwu (1993), Assessment of quality of reported births and infant mortality in China 1990 census, *China 1990 Population Census – Papers for International Seminar*, edited by China Statistical Bureau, Beijing: China Statistical Publishing House, 195–201.

Wu, Cangping (1984), The characteristics of age structure of China's population. In *A Census of One Billion People*, edited and printed by the Population Census Office under the auspices of the China State Council and the Department of Population Statistics of the China State Statistical Bureau, 399–409.

7 Population Ageing and Old Age Security

Du Peng and Tu Ping

Since the 1970s, the implementation of family-planning policy and rapid decline in fertility in China has received a lot of attention both inside and outside that country. As one of the main results of the rapid fertility decline in a very short period, population ageing has emerged as a new challenge facing Chinese society. It is also projected that there will be an accelerated period of population ageing in the near future.

The ageing of the population is an inevitable process during the demographic transition. Like the socio-economic development, demographic transition, industrialization, modernization and urbanization that have taken place in the last several decades all over the world, population ageing has today become a global phenomenon. However, the developed and developing countries have proceeded at different tempos during their demographic transition and process of population ageing. In developed countries, the elderly population has already reached a fairly high proportion and has usually grown slowly. But in most of the developing countries the proportion of elderly is still low, but will increase much faster than in the developed countries due to a rapid decline of fertility and mortality over a very short period of time. Generally speaking, all developed countries and most developing countries are facing an unprecedented demographic shift, namely, population ageing. China is no exception.

The Trend of Population Ageing in China

There has been significant change in the size and age structure of China's population since the 1950s. Owing to a rapid decline in infant and child mortality and to high fertility, the proportion of youth population (aged 0–14) among the total increased, and the proportion of elderly population (aged 65 and over) decreased, in the 1950s and 1960s, leading to a rejuvenation of the population as a whole. With the big drop in fertility after the 1970s, the proportion of youth population declined rapidly, from 40.7 per cent in 1964 to 33.6 per cent in 1982 and 27.6 per cent in 1990. Meanwhile, the proportion of elderly population increased steadily, from 3.6 per cent in 1964

Table 7.1 China's population size and age structure, 1953–2050

Years	Total population (millions)	Population aged 65 and over (millions)	Population aged 0–14 (%)	Population aged 65 and over (%)	Total dependent population (%)
1953	583	25.0	36.3	4.4	40.7
1964	691	24.7	40.7	3.6	44.3
1982	1,004	39.3	33.6	4.9	38.5
1990	1,131	62.9	27.6	5.6	33.2
2000	1,304	87.4	27.1	6.7	33.7
2010	1,340	107.9	22.4	8.1	30.4
2020	1,483	160.8	19.0	10.9	29.9
2030	1,519	223.9	17.4	14.7	32.2
2040	1,509	298.9	15.4	19.8	35.2
2050	1,462	306.8	14.7	20.9	35.6

Sources: *Almanac of China's Population* (1993, p. 153); Du, 1994, pp. 55–7, 80; Tu Ping, 1996.
Notes: Figures for 1953–90 are observed values; figures for 2000–50 are projected values; Taiwan, Hong Kong and Macao are not included in the statistics and discussions.

to 4.9 per cent in 1982 and 5.6 per cent in 1990. The age structure of China's population changed from that of a youth population to that of an adult population (Table 7.1).

China has the largest elderly population, as well as the largest total population, in the world. According to the 1990 national population census data, there were 1.13 billion in the total population in 1990, among them about 97.3 million people aged 60 and over, accounting for 8.57 per cent of the total; people aged 65 and over accounted for 5.57 per cent (table 7.1). The rapid demographic transition in China and particularly the rapid decline in the fertility rate, have resulted in an accelerated ageing process for China's population. By the end of 1996, China's total population exceeded 1.22 billion; there were more than 120 million people aged 60 and over and 78 million people aged 65 and over, the people aged 65 and over accounting for 6.4 per cent of the total population.

It is projected[1] that the proportion of the elderly aged 60 and over will increase to 22 per cent in 2030 when those who were born during the two baby booms (1950s and 1960s) reach their advanced years, and the proportion of the population aged 65 and over will increase to 15 per cent. The number of elderly people aged 65 and over will reach 224 million by 2030 and more than 300 million by 2050. So the ageing issue is of great importance for China and its socio-economic development, especially with regard to concerns about the increase in the dependency ratio (table 7.1). This trend is unchangeable because all those who will be the elderly in the middle of next century have already been born, and the change of the number of elderly population will only be affected by the mortality rate.

Compared with developed countries and even other developing countries, China's population ageing has some unique characteristics.

1. *Unprecedented speed.* Fertility in China declined further after the stagnation of the 1980s and reached replacement level 8 years earlier than the government's original population plan had projected (Zhang and Jiang, 1995). Consequently, it took less than the 20 years it took in Japan for China's elderly population aged 65 and over to increase from 5 per cent to 7 per cent. The time interval between 7 per cent to 14 per cent is also very likely to be less than the 26 years it took for the same increase to occur in Japan. Japan is the country with the most rapid pace of population ageing observed so far. For the proportion of population aged 65 and over to increase from 7 per cent to 14 per cent, it took France 115 years, Sweden 85 years, Germany and the United Kingdom 45 years (Du, 1994). Therefore, China's population will age at a speed unprecedented in the world, if the currently predicted trends in fertility and mortality hold.

2. *Early arrival of an aged population.* The developed countries did not have a severely aged population until they had completed their modernization processes and reached a high level of socio-economic development so that they had sufficient resources to establish and support an adequate social security and service system for the elderly. In contrast, China's rapid demographic transition started in the early stage of its modernization process at a relatively low level of socio-economic development and will be completed in a much shorter period of time. It is certain that China will face a severely aged population before it has sufficient time and resources to establish an adequate social security and service system for the elderly.

3. *Fluctuations in the total dependency ratio.* Due to the rapid decline in fertility, the proportion of youth population will decline rapidly, and the proportion of the elderly population will increase at first slowly and then at an accelerated rate. As a result, the proportion of the dependent population among the total population will first decline significantly, reaching a trough in the 2010–20 period. It will then increase steadily. Therefore, the country will first experience a period when there is an abundant labour supply and light dependency burden, and then a period with high dependency burden. Due to the fact that the average consumption of an elderly person is about twice of that of a child (Yu, 1994), the increase in the actual dependent burden will come much earlier and reach a higher level than currently estimated.

4.   *Strong influence of the government's fertility policy and its imple-
     mentation on the ageing process.* It is expected that mortality in China
     will decline steadily but at a slower pace in the future and fertility
     will be the main factor determining the future process of population
     ageing. Therefore, the government's fertility policy and its imple-
     mentation will be very important in determining the pace and timing
     of population ageing in the country.

The above analysis shows that China is going to face a serious conflict
between control of its population size and optimization of its population
age structure. In order to control its population growth effectively and
achieve zero population growth with a smaller population total, China
needs to keep its fertility low. This will lead to a sharp decline in its child
dependency ratio and total dependency ratio in the short run, but will eventually lead to an extremely high elderly dependency ratio and total dependency ratio in the long run. On the other hand, higher fertility is needed to
produce a relatively good population age structure in the future and to
avoid too rapid population ageing. But that will lead to a growing population for a long period of time, a postponed arrival of zero and negative
population growth, and great pressure on the country's environment and
resources.

The Decline of Fertility and Mortality and their Effects on Population Ageing in China

To analyse the effects of fertility and mortality decline on the process of
population ageing in China, the method of comparative population projections is used (United Nations, 1988). Several population projections are prepared with the same initial population and different assumptions of fertility
and mortality trends, and differences in the age structure of projected populations are broken down into fertility effects and mortality effects. A set of
four comparative forward projections has been prepared for each of two
periods: 1950–90 and 1990–2030.

The results demonstrate that both fertility and mortality are the major
determinants of population ageing in China. The fertility decline is in
general a much more important factor than mortality decline (table 7.2).
However, as the total fertility rate (TFR) has already reached a relatively
low level now, the importance of further changes in mortality should not be
underestimated in the process of population ageing in China in the future
(Du, 1994).

Regional and Rural–Urban Differences in Population Ageing

Although China has not yet ranked among countries with an old population pattern, we can not say that the problem of population ageing does not

Table 7.2 Breakdown of changes in the proportion of population, China, 1950–1990 and 1990–2030

	1950–1990	1990–2030
Under age 15		
Initial proportion	33.5	27.6
Proportion at the end	27.6	18.0
Absolute changes	−5.9	−9.5
Fertility effect	−19.5	−4.1
Mortality effect	3.0	−0.3
Effect of the initial age distribution	10.6	−5.1
Aged 60 and over		
Initial proportion	7.5	8.6
Proportion at the end	8.6	21.9
Absolute change	1.1	13.3
Fertility effect	2.7	2.1
Mortality effect	0.8	1.7
Effect of the initial age distribution	−2.4	9.5

Source: Du (1994) *The Process of Population Ageing in China*, p. 114.

exist in the country. Regional differences have been found in socio-economic development and in the in age structure of the population; there are corresponding differences in the level of population ageing between rural and urban areas, and among provinces, autonomous regions and municipalities.

Due mainly to significant regional and rural–urban differences in fertility and mortality, as well as in geography, climate and socio-economic development, Chinese elderly people enjoy very different living conditions, welfare, medical and health care and pension systems depending on where they live.

1. *Urban–rural difference*

According to the 1990 census data, of the 97.25 million old people aged 60 and over in the country, 18.38 million lived in cities, 6.02 million lived in towns and 72.85 million in counties, accounting for 18.9 per cent, 6.2 per cent and 74.9 per cent of the total elderly population respectively. In other words, a quarter of the elderly population lived in urban areas while three quarters lived in rural areas, which was consistent with the level of urbanization of the population as a whole. The distributions in 1982 were 14.0 per cent, 5.2 per cent and 80.8 per cent.

It is usually thought that the proportion of elderly people in cities is higher than that in counties, but the available data show that the rural area has a higher proportion of elderly population aged 60 and over than the urban area.

In 1982 people aged 60 and over accounted for 7.4 per cent, 6.4 per cent and 7.8 per cent, whereas people aged 65 and over accounted for 4.67 per cent, 4.22 per cent and 5.01 per cent of the population in cities, towns and counties, respectively. In 1990 the proportions of people aged 60 and over increased to 8.6 per cent, 7.2 per cent and 8.7 per cent respectively. The ageing process took place in urban areas earlier than in rural areas, but the rural areas have a higher proportion of elderly people than the urban areas. This can be explained by the fact that people of working age migrate from rural areas to cities and towns, and particularly to towns located near rural areas.

2. *Inter-provincial difference*

Figure 7.1 illustrates the differences in the proportion of elderly people among all the provinces, autonomous regions and municipalities in 1995. It indicates that, in 1995, Shanghai had the highest level of ageing in China with the proportion of elderly people aged 60 and over being 16.7 per cent (elderly aged 65 and over, 11.4 per cent). On the other hand, Qinghai province had a lowest level of ageing with the proportion of the elderly (60 and over) being only 6.3 per cent (elderly aged 65, 3.6 per cent). There exist large differences between the regions in China. As regards the proportion of the population aged 0–14 years old, the highest and lowest figures in 1995 were for Tibet and Shanghai, being 35.4 per cent and 17.1 per cent, respectively.

In accordance with the level of ageing, China can be divided into three regions. The first is the eastern coastal region that is characterized by a high population density and high level of ageing. The second is the middle part of the country where there is middle-level ageing. The third region consists of the provinces in the northern, western and southern parts of China, where most minority nationalities live. Generally speaking, the last region has a less developed economy and is commonly sparsely populated. Moreover, the fertility level in this region has been relatively high in the past dozen years resulting in a young age structure and a low level of ageing. Since the total fertility rates in three big municipalities, namely, Shanghai, Beijing and Tianjin, have long been standing at 1.3–1.4 which are among the lowest fertility levels in the world, they have already encountered a serious ageing problem. The situation is most alarming in Shanghai, as the growth rate of its native population has been negative in the last five consecutive years. By the end of 1998, the elderly population aged 65 and over already accounted for 12 per cent of the city's total population, a figure similar to many developed countries.

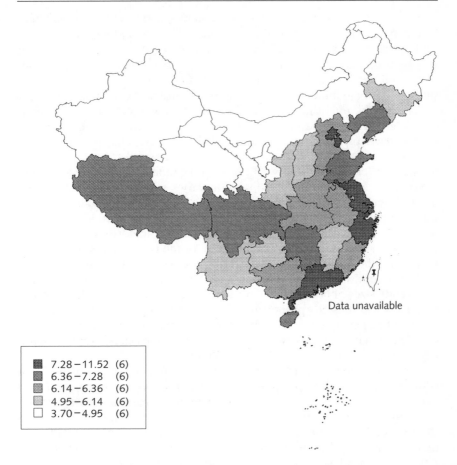

Data unavailable

■	7.28−11.52	(6)
■	6.36−7.28	(6)
■	6.14−6.36	(6)
▢	4.95−6.14	(6)
☐	3.70−4.95	(6)

Figure 7.1   Proportion of people aged 65 and over, China, 1994

Therefore, taking China as a whole, it will still take a dozen or so years before the country faces a serious ageing challenge, but the large cities and east coast provinces with very low fertility and mortality rates have already, in advance, acquired an old-type population. As a result, the pressure of population ageing in these areas will increase quickly and a number of problems connected with low fertility level are emerging.

Ethnic Differences on Ageing

There are 56 recognized nationalities in China. As the size of population varies with nationalities, the number and proportion of the elderly people of each nationality also varies greatly.

The 1990 national population census revealed that the population belonging to minority nationalities numbered 91.2 million, accounting for 8.04 per cent of the country's total, the remaining 91.96 per cent being of Han nationality. The Zhuang nationality is the largest population subgroup among all national minorities, recording more than 15.41 million. A few smallest national minorities have only two thousand or so people each.

The distribution of nationalities in China follows an obvious geographical pattern. The people of Han nationality live mostly in the eastern and central parts of the country, while people from national minorities live mainly in the western, southern and remote regions. As the result of differences in fertility, mortality and in socio-economic development among regions, the age structure of population varies with nationalities; so does the ageing process. In 1982 the proportion of Han people aged 60 and over was 7.69 per cent, which is slightly higher than the national average (7.64 per cent). The figure for all the other 55 national minorities put together was 6.88 per cent, a lower proportion than for the Han. Among the national minorities, the Russians had the largest proportion of the elderly people, 11.48 per cent, while the Elunchun had the lowest, 2.49 per cent. But the population sizes of both these two nationalities are very small. As for other national minorities, the proportion of the elderly population ranges from 5 per cent to 8 per cent. It is thus clear that the size and proportion of China's elderly population are overwhelmingly determined by changes in the population dynamics of people of Han nationality.

The proportion of elderly population aged 60 and over for China as a whole increased to 8.57 per cent in 1990. The proportion of people of Han nationality was 8.67 per cent, still higher than the national average, while the figure for all 55 national minorities was 6.94 per cent. Compared with the 1982 census, the proportion of the elderly of both Han nationality and minority nationalities had increased, but the upward trend in the Han population was much faster than in the minority nationalities.

Among minority nationalities, large differences still exist. There were 19 nationalities with populations of more than 1 million in 1990. Ten of them witnessed an increase in the elderly proportion during the period 1982–90, while nine experienced a decrease. Not every nationality in China, therefore, is ageing. As most nationalities with a decrease in their proportion of elderly people are small, their influence on the overall situation is quite limited. On the whole the elderly proportion of China's minority population is increasing.

The low proportion of the elderly among minority nationalities can be attributed mainly to their higher fertility levels and young age structures. The Chinese government has given priority to the prosperity of minority populations, therefore different population policies and programmes have been implemented for the majority Han population and the national minorities. In general, population control in minority regions is much more flexible[2] and the family-planning programme has to be adapted to the actual conditions of each national minority. On the other hand, minority-concentration areas are relatively less developed, and population reproduction there is at the initial stage of transition. In other words, the birth rate there remains relatively high while the death rate is starting to drop. As these differences will not vanish in a short time, the age structure of the national minority population will continue to be younger than that of the Han population for quite a long period of time in the future.

The Change in the Family Life of Elderly People in China

In the course of the decline of Chinese fertility and mortality, and the ageing of the population, changes are taking place in the structure of family types and in the size of families (for a detailed discussion, see chapter 9). The nuclear family is gradually taking the place of the extended family as the main form of family, accounting for 67 per cent of the total of families in 1995. Three-and-more-generation lineal families are the next most important family type after the nuclear family, accounting for 17 per cent of the total. Small families consisting only of one couple or a single person are, however, becoming increasingly common for old people.

At the same time, the average size of Chinese family households is tending to shrink. Three- or four-person family households have replaced four- or five-person households as the dominant type in China. Whereas the average family size was 4.43 persons in 1982, it had been reduced to 3.96 in 1990, and had further decreased to 3.70 by 1995.

The drop in the birth rate has reduced the number of children in each family who can shoulder the responsibility of supporting their aged parents. As the Chinese family is the main base for providing care for the aged, young people nowadays have a heavier burden of old-age care. As the population ages, families without children or with only one child will increase in proportion. Therefore, more and more old people will have no children or only one child to provide for them.

The majority of the elderly are living with their children in two- or three-generation families. In 1990 the average size of the family households of old people aged 60 and above was 4.11 persons, higher than the national average of 3.96 persons. Among the elderly, 21.5 per cent lived in two-person families, 17.9 per cent in five-person families and 25.6 per cent in six-and-more-person families.

The marital status of old people has some impact on their family size. In 1990 50 per cent of unmarried and 42 per cent of divorced old people lived alone. Widows and widowers generally lived together with their children: only 17 per cent of them lived separately, while the remaining 83 per cent lived in two-and-more-person families. It is clear that unmarried and divorced single old people are more in need of care and help from society. A computer simulation shows that women aged 60 and over will eventually constitute 36 per cent of the total female population in urban areas, with 40 per cent of them living alone, if urban fertility remains at the level achieved in 1986 for a long period of time (Zeng, 1991).

The average household size of Chinese old people in 1990 was 3.73, 3.75 and 4.20 persons in cities, towns and counties respectively. As more of the rural elderly live in families containing 3 and more generations, the family size of the old people is larger in the countryside than in cities. The proportion of the old people living in families containing four or more people was 52.2 per cent in cities, 50.6 per cent in towns and 58.6 per cent in counties.

In 1995 the number of Chinese people aged 60 and over surpassed 110 million. The life of the aged population and their marital and family status are increasingly becoming a focus of attention for society. In China, where the family is the principal means of providing care for the aged, the marital and family status of the old people has a direct bearing on their quality of life in their twilight years.

As regards the marital status of old people, in 1990, old people with spouses were in the majority; the proportion being markedly higher than in 1982. The proportion of the old with another marital status was on the decrease; among them, the proportion of widowed dropped by the biggest margin. The above-mentioned changes in marital status have a positive influence on the life of the aged. It is significant that, although the proportion of single, widowed and divorced old people decreased somewhat, the total number of old people in these marital categories had increased because of the steady growth of the aged population as a whole. Nowadays most unmarried and divorced old people are males living in the countryside, and the proportion is still growing.

With advancing age, the proportion of the old people living in families containing three and more generations continues to grow. In terms of urban-rural difference, more rural old people live with their children than their urban counterparts. This reflects the great difference in economic level and social attitudes between urban and rural areas. In terms of sex difference, more old females than old males live with their children. This has much to do with the gender difference in income and social status of old people. When old people live with their children there are at least two and generally more generations under one roof. The family size is, therefore, comparatively large.

Social Support and Family Support of the Elderly

It is stipulated in the Law of the People's Republic of China on the Protection of the Rights and Interests of the Elderly (which came into force as of 1 October 1996) that: 'the elderly shall be provided for mainly by their families, and their family members shall care for and look after them.' In the Marriage Law it is stipulated that: 'children are in duty bound to support parents'.

For a long time in China, family support of the elderly was considered as the only option for aged people. As Chinese families become smaller and much more mobile than ever before, the existing family-based care system for the elderly has been under great pressure. The state and the social security system should try to take on more economic responsibilities for the family. However, although social security for the elderly has been widely acknowledged as a crucial social programme, government social welfare expenditure on the retired in 1992 only constituted 3.5 per cent of the gross national income; only 25 per cent of the country's elderly population were covered by the current social security system. Meanwhile, the average savings of the elderly population are rather low compared with those currently at work (Yu, 1994).

Therefore it is rational for people, particularly in rural areas, to want to have more children – especially sons – to take care of them when they become old. The traditional thinking based on a thousand-year feudal tradition still affects many member of the rural masses, for them the saying of 'the more children, the greater happiness' remains valid. The promotion of 'one couple one child' since the 1970s raises the question of who will support the parents when they become old (Wu and Du, 1994).

The issue of the only child is much more acute in cities than in the countryside. Up to now, most rural women still have two or even three children, but urban women, particularly those in metropolitan cities, predominately have only one child. The proportion of families with only one child was as high as 70 per cent in 1990 in Shanghai, and 56 per cent in Beijing. If only the inner city itself is taken into consideration and the suburbs are excluded, the rates are even higher.

Only children are often labelled 'little emperors'. Some people are worried that China will bring up a generation of spoilt children, selfish individuals, unable to co-operate with others and unwilling to take care of their old parents and relatives etc. Some people argue that the Chinese family is in a stage of transition from traditional to modern; children are still seen as important as heirs to continue the family line, or as security for the future, but parents are becoming more child-centred than before (Wu and Du, 1994).

In the cities, as only-child families increase, many people raise the problem of the 'four, two, one' pattern of family support in the future. This is to say that while the first generation of couples who have only one child have a very light dependency burden, their single child and his or her spouse will have to support four parents when the latter reach retirement age. As life span is prolonged, many may have to take care of several great-grandparents, even though they live in a separate residence.

Although family support is very important, under conditions of decreasing fertility and the nuclearization of the family to depend solely on family support is definitely not enough. In many cases taking care of elderly may be too heavy a burden for their children and family members to carry even if they want to do their best. The reform of the social security and social service system is one of the priorities in China today. The guideline suggested is that the support of elderly should be a joint effort, that is, the state, community and family should share the responsibility. The crucial challenge with respect to population ageing will occur at the beginning of the twenty-first century when the baby-boom generation reach old age. The baby-bust generation will have a hard time supporting them unless preparations are made beforehand.

Families provide the elderly not only with economic security but also with emotional care that outsiders cannot replace. But, on the other hand, the world is changing; industrialization, migration and the modernization of lifestyles, all result in large-scale changes in attitude towards family support for the old.

In the 1992 Beijing Multidimensional Longitudinal Survey, most of the elderly respondents (63.8 per cent) agreed that to have a son is to protect one's old age; only 21.4 per cent disagreed (Myers and Du, 1996). And almost half of the elderly interviewed agreed that young people today had less respect to the elderly. In the process of rapid modernization, the perceptions of the younger generation are changing dramatically. For example, many more educated young couples prefer to have smaller-sized nuclear families. They prefer not to live with their parents to avoid family conflicts. They prefer to send their children to kindergartens to get a better pre-school education instead of keeping them at home with their grandparents and in case grandparents spoil them. In addition, most of the young generation think the experiences of the elderly are of no relevance to modern society, while the old generation still adheres to the stereotypes of the generations that preceded them. These changes are manifested in the answers of elderly respondents in the 1992 survey. As a result, the generation gap is tending to widen.

By 1994 even more of the elderly believed that to have a son is to protect one's old age, the proportion agreeing with this statement increasing from 63.8 per cent in 1992 to 73.6 per cent, a ten-point change. It seems that as the elderly age further, they become more and more convinced that their sons are the only security they have for later life.

Conclusion

The above analysis shows the great challenge that China will have to face in controlling its population growth and avoiding too rapid population ageing. It is time for us to address the negative impacts of a strict fertility control and take proper action (Gu and Mu, 1994). In addition to those discussed in the previous sections, the following concerns are not unfounded either. The cost of childbearing and child-rearing is mainly shouldered by the family. The government pays only part of the cost of educating children, and now its relative share is declining with the development of market economy. Therefore, the direct savings in government expenditure from reducing fertility are relatively limited, and its main benefits lie in the alleviation of the pressure on the environment and resources caused by a rapid population growth. The potential savings for the family from a reduced number of children will also be partly offset by the increase in the unit cost per child due to a loss of economy of scale under a low fertility regime. Furthermore, the reduction in the number of dependent children in the family often leads to an increase in the level of current consumption instead of a significant increase in investment or long-term savings, when investment opportunities are poor and uncertain, financial and insurance institutions are underdeveloped, and there is severe inflation. Therefore, while strict fertility control is beneficial to environment protection and resource conservation, it tends to transfer some of the dependency burden of the current generation to the next generation and cause unexpected problems for sustainable development. The direct economic return from such a policy is likely to be more uncertain and smaller than expected.

The recent population projections for China (Zeng, 1994; Li, 1995) indicate that: China's total population will start to decline around 2030 when it reaches 1.5 billion; the proportion of the elderly (65 and over) will reach 7 per cent in year 2000, 14 per cent in 2025 and 23 per cent in 2050; and the total dependency ratio will fluctuate between 40 and 70 if a smooth and gradual transition to a two-child policy is initiated now. Such a policy option is very close to people's fertility desires, and will still assure the achievement of the population target set by the government (keeping China's total population below 1.3 billion by year 2000 and below 1.6 billion throughout the 21st century). It may provide the country with some time to establish and improve its social security and service system to support its elderly population before they constitute a very high proportion of the total population. It will also lead to a more balanced distribution of the dependency burden among different generations. Therefore, a two-child policy provides an alternative that gives a reasonable balance between controlling population size and optimizing population and family structure. Such an option merits further serious study.

Notes

1. The projection is based on the following assumptions: that the total fertility rate will decline from 2.31 in 1990 to 1.8 by the year 2000 and remain at that level up to 2050; that male life expectancy at birth will improve from 67.58 years in 1990 to 75 years to 75 years by 2050 and female life expectancy at birth will also improve from 70.91 years in 1990 to 80 years by 2050.
2. See discussions in chapter 5.

References

Du, Peng (1994), *The Process of Population Ageing in China* (in Chinese). Beijing: Press of the People's University of China.

Gu, Baochang and Mu Guangzong (1994), A reconsideration of population problems in China, *Population Research*, 1994 (5), 2–10 (in Chinese).

Institute of Population Research, Chinese Academy of Social Sciences (1993), *Almanac of China's Population* (in Chinese). Beijing: Economic Management Press of China.

Li, Jianxing (1995), *A Study on the Smooth Transition of the Fertility Policy in the 21st Century for Rural China* (in Chinese). Ph.D. thesis, Institute of Population Research, Peking University.

Myers, George C. and Du, Peng (1996), *Living arrangements among Chinese older persons*. Paper presented at the annual conference of American Association of Gerontology, Washington DC.

Population Census Office under the State Council and the Department of Population Statistics of the State Statistical Bureau (1993), *Tabulation of the 1990 Population Census of the People's Republic of China* (in Chinese). Beijing: China Statistical Press.

Tu, Ping (1995), On the issues of population ageing and population control in China, *Population Sciences of China*, 1995 (6), 61–70 (in Chinese).

United Nations, Department of International Economic and Social Affairs (1988), *Economic and Social Implications of Population Ageing*. ST/ESA/SER. R/85, New York.

Wu, Cangping and Du, Peng (1994), The demographic aspects of population ageing in China: Social and economic implications. In *The Demographic Aspects of Population Ageing and its Implications for Socio-Economic Development, Policies and Plans,* monograph organized by INIA and CICRED, Malta.

Yu, Xuejun (1994), *An Economic Study of Population Ageing in China* (in Chinese). Ph.D. thesis, Institute of Population Research, Peking University.

Zeng, Yi (1991), *Family Dynamics in China*. Madison: the University of Wisconsin Press.

Zeng, Yi (1991), *Project Report on the Mid and Long Term Population Projection* (in Chinese). Working paper, Institute of Population Research, Peking University.

Zeng, Yi and Vaupel, James (1991), Several problems concerning Chinese future population developmental trends in China, *Social Sciences in China* (Chinese edition), 1991 (3), 3–17.

Zhang, Lingguang and Jiang, Zhenghua (1995), Perspective of population in China, *Population Science of China*, 1995 (3), 1–7 (in Chinese).

8 Marriage Patterns in Contemporary China

Zeng Yi

Since the onset of economic reform in China at the end of the 1970s, the Chinese economy has developed rapidly and the standard of living of Chinese people has been improved and changed substantially. These changes have been accompanied by significant demographic changes, including changes in marriage patterns. The first section of this chapter briefly addresses the data sources. The subsequent sections will deal with the recent trends in marriage timing, age pattern, and intensity since 1980, respectively.

The data presented here are mainly from China's two-per-thousand fertility and contraceptive survey conducted in July 1988 by the State Family-Planning Commission. In this survey, 2.15 million persons were actually surveyed. They included 459,269 women aged 15 to 57 who had been married at least once or were still married and who were interviewed in detail on 67 items including their marital status and date of marriage. The data for 1990 is from China's fourth census conducted in July 1990. To give an overview of the trends of Chinese marriage in the past 50 years, also included below are some figures of first marriage rates before 1981, derived from the one-per-thousand fertility survey conducted in 1982. The 1982 survey also covered the whole country with a sample fraction of one per thousand of the total population. It also collected detailed information on marital status and date of first marriage. The high quality of the data from the 1988 survey, the 1982 survey and the 1990 census is widely recognized.

To discuss the changes in age at first marriage after 1990, relevant data derived from the 1992 fertility survey conducted by the State Family-Planning Commission has been used. The sample size of the 1992 survey was 385,000 persons, and among them 114,000 were women of reproductive ages from 15 to 49. The average sampling proportion is 0.35 per thousand. The survey also collected background information about marriage. It has been discovered that the 1992 survey seriously underreported recent births (see, for example, Zeng Yi, 1996). Although we do not rule out the possibility of a misreporting of marital status by those married below the legal minimum age of marriage (20 years old for women and 22 for men), we trust the 1992 survey's data quality on marital status and date of first marriage to be generally reasonable, because the penalties for illegal early marriage are much smaller than for out-of-plan births, and marriage is generally not so strictly included in the 'responsibility system'.[1]

Continued Universality of Marriage

Coale (1984) estimated the proportion of women ever-married aged between 15 and 35 for the period 1950 to 1982 from the one-per-thousand fertility survey data. For every cohort that had reached the age of 35 by 1982, more than 97.5 per cent of the women were married by the age of 35, and for most of the cohorts the total exceeded 99 per cent. Was this universality of marriage continued in the 1980s? We can easily extend Coale's estimate to 1988 based on the two-per-thousand fertility and contraceptive survey data, but it will not answer the above question completely since we do not know the behaviour of people younger than age 35 in 1988. The life table approach handles this problem adequately.[2] table 8.1 and figure 8.1 show that about 99 per cent of Chinese women have been married at least once according to the results of the life tables for the 1980s. This provides evidence to show that the pattern of nearly universal marriage has been continued into the 1980s, probably the result of the continued cultural value attached to marriage in China.

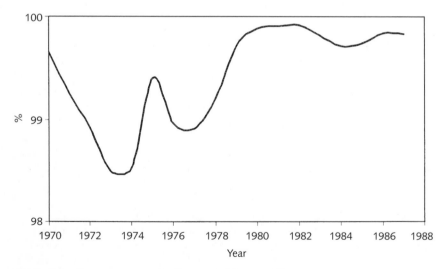

Figure 8.1 Percentage eventually married by age 50

Table 8.1 Some major marital indices

Year	% eventually married by age 50 (a)	Female mean age at first marriage (b)
1970	99.65	
1971	99.25	
1972	98.92	
1973	98.49	
1974	98.53	
1975	99.40	
1976	98.96	
1977	98.91	
1978	99.21	
1979	99.72	
1980	99.88	22.84
1981	99.90	22.67
1982	99.91	22.41
1983	99.81	22.35
1984	99.71	22.28
1985	99.73	22.24
1986	99.83	22.13
1987	99.82	22.16
1987–90		22.00
1992		22.60

Note:
(a) Female intensity of first marriage, measured by the life table approach.
(b) The figures for the period of 1980–7 were derived from the 1988 two-per-thousand fertility and contraceptive survey; the figure for 1987–90 was estimated from the 1987 one per cent population survey and 1990 census data; the figure for 1992 was derived from the 1992 fertility survey.

Changing Age At First Marriage

Age at first marriage shifted upward approximately four years between 1950 and 1980 and most of this increase occurred in the 1970s. The substantial increase in age at first marriage in the 1970s was mainly due to the impact of the emphasis put on delayed marriage in the Chinese family-planning programme of '*Wan, Xi, Shao*' (Coale, 1984; Banister, 1987; Zeng et al., 1985, 1994).

Since 1980, the age at marriage has fallen. The range of unconditional first-marriage rates[3] higher than 10 per cent went down from ages 21–5 in

1979 to 20–24 after 1980. Age-specific first-marriage probability among never-married women reveals the same trend from another angle. For example, if we follow the 10 per cent level probability of first marriage – above this level, more than a tenth of unmarried women get married each year. The 10 per cent level can, therefore, be viewed as representing the start of a period of intense marriage activity during which a large majority of Chinese females marry. The starting point for the 10 per cent level stood at age 19 in 1970, went up to age 22 in 1974–9, and went down to age 20 after 1980. The starting point for the 20 per cent level went down from age 25 before 1979 to age 23 in 1981–85 and down again to age 22 in 1986–7. The female mean age at first marriage decreased from 23.0 in 1979 to 22.4 in 1982 and to about 22.2 in 1987, and 22.0 in 1987–90[4] (see table 8.1 and figure 8.2).

The first quartile, median, and third quartile age at first marriage fell between 1982 and 1990 (see table 8.2). The male median age fell from 24.9 in 1982 to 23.6 in 1990. The median age at first marriage for females decreased from 22.9 in 1982 to 22.3 in 1990.

The decrease in age at first marriage at the beginning of the 1980s was mainly due to the relaxation by the New Marriage Law, which took effect officially on 1 January 1981, of the restrictions on age at first marriage enforced in the 1970s. Although the 1981 New Marriage Law shifted the minimum legal age for marriage from 20 for males and 18 for females (specified by the 1953 Marriage Law, but not in force in 1970s), to 22 for males and 20 for females, the administratively controlled minimum age at first marriage (mid-20s for females and late 20s for males) lost its ground for implementation.

Table 8.2 The median and the quartiles of age at first marriage

	Year	Male			Female		
		1st quartile	Median	3rd quartile	1st quartile	Median	3rd quartile
Whole China	1982	23.00	24.90	27.05	21.01	22.89	24.66
Whole China	1987	22.15	23.67	25.75	20.73	22.35	23.97
Whole China	1990	22.09	23.63	25.64	20.77	22.27	23.88
City	1990	23.33	25.04	27.10	21.89	23.41	24.90
Town	1990	22.90	24.54	26.41	21.50	23.08	24.67
Rural areas	1990	21.71	23.20	24.96	20.53	21.90	23.50

Note: The 1982, 1987 and 1990 figures were derived from the 1982 census, the 1987 one per cent population survey and 1990 census, respectively.

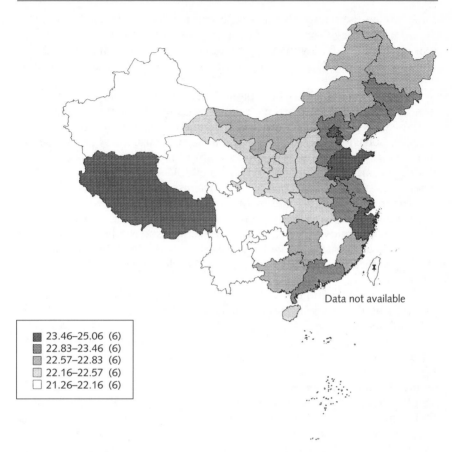

Data not available

23.46–25.06 (6)
22.83–23.46 (6)
22.57–22.83 (6)
22.16–22.57 (6)
21.26–22.16 (6)

Figure 8.2 Mean age at first marriage for women, China, 1994

The Chinese economy has been developing quickly and living standards have been improving substantially since 1980. Thus, the fact that the age at first marriage in China continued to fall gradually after 1984 is an interesting phenomenon because the theoretical expectation is of a positive relationship between age at marriage and economic level. Several explanations may shed light on this issue. First, from the beginning of the 1980s, many government agencies and the public media increasingly emphasized the important function of legal systems, and marriage registration came more closely under the control of the departments of civil affairs rather than family-planning programme offices. The departments of civil affairs emphasize the citizen's rights to choose when to marry under the New Marriage

Law, making the family-planning offices in many areas less able to implement late marriage for birth-control purposes. This has been a source of tension between family-planning programme leaders trying to promote late marriage, and those working in civil affairs who argue that they must maintain the law.

Second, even in places where family-planning offices were still in a position to implement late-marriage regulations, the administrative decentralization since economic reform has weakened governmental control over people's behaviour including age at marriage.[5] Third, many people whose income increased in 1980s were able to afford their own or their children's marriages earlier than they would have in the 1970s. A fourth factor is that, given that brides usually marry into grooms' households and almost all females in China participate in the labour force, earlier marriage (early childbearing usually follows) is one way of increasing the labour force in grooms' households, especially under the rural responsibility system, which assigned land to each household after 1980. This is one of the rationales for young people in rural areas to get married earlier in the 1980s than in the 1970s.

The fifth explanation is a demographic one. As Tian (1991) correctly pointed out, the problem of the 'marriage market', caused by the suddenly tremendous decrease in fertility in the period 1959–61 when the total fertility rate was 3.87 in contrast to about 6 immediately before and after this period, is one of the reasons why age at marriage fell in 1980s.[6] Because of the extremely low fertility between 1959–61, the number of women born in this period and aged in their twenties in the 1980s is much smaller than that of males born between 1956–8 and about three years their seniors. Husbands are usually two to three years older than their wives in China. Some of the males born in 1956–8 were not able to find a spouse two to three years younger than themselves, so they sought females who were four, five or even more years younger. This may have pushed more young women to marry and caused the decrease of age at marriage for females. On the other hand, the number of women born in 1962–4 was much larger than the number of men born in 1959–61, and marriage pressure from the female side might have pushed more men to marry at a younger age.

According to the 1992 survey, the female average age at first marriage rose from 21.9 in 1987 to 22.0 in 1990, and to 22.6 in 1992. The percentage of the ever-married at ages 19 and 20 decreased from 30.1 and 47.0 per cent in 1987 to 29.2 and 44.0 per cent in 1990, and to 18.0 and 34.4 per cent in 1992. The percentage ever-married at other ages before age 25 also decreased significantly. Rapid economic development may have contributed to the recent increase in age at first marriage. Many young peasants who left their villages to work in urban areas or engage in private enterprise may have delayed marriage for practical reasons. Stronger implementation of the policy promoting late marriage has also contributed to the increase in age at first marriage. The fact that, while continuing the free distribution to married couples, local medical shops have started selling contraceptives in recent years has helped unmarried couples who engage in sexual activity to avoid early marriage as a consequence of pregnancy.

Table 8.2 also reveals the significant differences in the timing of first marriage between Chinese cities, towns and rural areas. For example, the male median age at first marriage in rural areas in 1990 was 1.84 and 1.34 years younger than that in the cities and towns, the female median age in rural areas was 1.51 and 1.18 years younger than that of city and town counterparts, respectively. This is not a surprising phenomenon, given the existing large differences in socio-economic conditions, degree of modernization as well as population policies between the Chinese rural and urban areas.

The age of marriage is in general higher in china's east coastal provinces, and is gradually reduced towards the inland provinces. This pattern could be attributed to a variety of factors. Generally speaking, the east coastal region of China is more advanced by Chinese standards, in terms of economic and social development indicators. While inland China is a more rural dominated and economically backward region. The gap, both in marriage pattern and socio-economic development, will most likely be maintained for a while in the near future.

Concentration of First Marriage in a Narrow Age Interval

One striking feature of marriage in China is that it is concentrated in a narrow age interval. The age above which the unconditional rates of first marriage are less than 0.5 per cent stood at 27 in the 1950s, went up to 28 in the 1960s, to 28–9 in the 1970s and to 29–30 in the 1980s – which means that virtually no women (less than 0.5 per cent) got married after 26–8 in 1950–69, and after 28–30 in the 1970s and the 1980s. The unconditional first marriage rates higher than 10 per cent concentrated at age 16 through to 19 in the early 1950s, age 18 through to 21 around 1970, and age 20 through to 24 after 1980. About 50–60 per cent of all marriages occurred within a narrow age bracket of 4–5 years. The Chinese cultural tradition is the background for such an impressive concentration of marriage. Chinese parents often hold the view that once age and other conditions of marriage under the given socio-economic situation are fulfilled, their sons and daughters should get married as quickly as possible. Since cohabitation outside of marriage and premarital sexual activities are culturally and legally prohibited, there is an emotional and practical pressure for eligible young men and women who fall in love to get married as soon as possible. Furthermore single men and women in their thirties are regarded by many as abnormal, and this social pressure also makes young people keen to marry quickly.

Summary

The major features of marriage patterns in recent decades in China can be summarized in four main points. (1) The Chinese pattern of universality of first marriage was evidently continued into the 1980s. (2) The age at first marriage increased dramatically in the 1970s, but decreased in 1980s on a modest scale, and increased again in the early 1990s. (3) There are significant differences in the timing of first marriage between Chinese cities, towns and rural areas. (4) The first marriage of Chinese women has been concentrated within a rather narrow age interval.

This chapter, has provided some qualitative speculations in terms of cultural, socio-economic, policy and demographic factors to try to explain the changes and persistence of marriage patterns in China in recent decades. However, these explanations are preliminary and descriptive. More detailed explanatory research and a regional differential analysis are needed for a full coverage of this subject.

Notes

1. The government has directed since 1991 that the number-one chief officers at provincial and all other lower levels should be personally responsible for meeting birth-control targets in their administrative areas. If the birth-control target is not reached, the number-one chief officer cannot be promoted or honoured and may even lose his or her job as leader. This is the so-called 'one-vote-down' (*yi piao fou jue*) campaign. The campaign has had some positive effects in strengthening the implementation of the family-planning programme. However, it has also had some negative effects, especially since it has increased the under-reporting of births in surveys after 1990. For the 'responsibility system', see also ch.5.
2. Some technical discussion is provided in Appendix 1.
3. The numerator of the unconditional rates is the number of women aged x who get married in year y and the denominator is the total number of women aged x in year y. The conditional probabilities are defined as the number of women aged x who get married in year y, divided by the number of women never married aged x at the beginning of year y.
4. Since there is no direct survey data on the timing of first marriage in the 1990 census, I estimated the average Mean Age at First Marriage during the period 1987–90 as 22.0 years old for women, based on the single-year age-specific proportion of single people as in mid-1987 and in 1990, derived from the 1987 one-per-cent survey and the 1990 census.
5. In some areas, the authority of the local government and Party may convince the department of civil affairs to co-operate with the family-planning office in implementing the late-marriage regulation, which is not provided for by the new marriage law.

6. The sudden tremendous decrease in fertility between 1959 and 1961 was caused by the Great Leap Forward and the natural calamities in the three years that followed.

References

Banister, J. (1987), *China's Changing Population*. Stanford: Stanford University Press.

Coale, A. (1984), *Rapid Population Change in China, 1952–1982*. Washington DC: National Academy Press.

Elandt-Johnson, R. C. and Johnson, N. L. (1980), *Survival Models and Data Analysis*. New York: John Wiley and Sons.

Tian, H. Yuan (1991), A probe to the causes of early marriage, *Journal of Population Science in China*, 2.

Zeng, Yi (1996), Is fertility in China at beginning of 1990s far below the replacement level? *Population Studies*, 50 (1).

Zeng, Yi, Vaupel, J. and Wang, Zhenglian (1994), Marriage and fertility in China: Recent trends, *Genus*, XLVIIIII (4).

Zeng, Yi, Vaupel, J. and Yashin, A. (1985) Marriage and fertility in China: A graphical analysis, *Population and Development Review*, 11 (4), 721–36.

Appendix I

We can compute the total first-marriage rates of each year by summing up the unconditional age-specific rates. For a cohort, these total rates can reveal the intensity of marriage, but we are not able to compute such total rates for cohorts younger than the minimum legal ages at marriage at the survey or census time. We can, of course, compute the period total first-marriage rate, which is, by definition, what would be the average number of first marriages per woman over her whole life in a hypothetical cohort subject to the given unconditional age-specific first marriage rates. However, when age at first marriage is declining, as it was in China in 1980s, the period total rates overestimates the real marriage intensity, and the total first marriage rate can exceed one. The reason is that the marriages of older and younger women, which would have occurred sequentially with constant mean age, overlap (Coale, 1984: 41). When age at first marriage is increasing, as happened in China in 1970s, the period total rate underestimates the real marriage intensity. For example, the total first mar-riage rates were as low as 0.64–0.68 in 1972–5 and as high as 1.27 in 1981, and 1.13 in 1986. Obviously, it is not correct to say, based on the values of those total rates, that the marriage level in 1972–5 implied that only 64–68 per cent of Chinese women would eventually marry. It is not even logical to say that marriage levels in 1981 and 1986 imply that Chinese women would, on average, have 1.27, and 1.13 first marriages. The life table approach does not solve entirely the problem of distortion to the true period marriage intensity. Nevertheless, the life table approach gives measurements of marriage intensity which are much more close to the true values and it never gives the non-logical results that total rates may give, such as first marriage intensity being greater than one when age at marriage is falling.

Another problem is that for women who are under the age of 49 and still in the status of never married, we do not know whether they will marry. This is the so-called 'right-censoring' problem. The life table approach can handle the right-censoring problem adequately, at least to a large extent. We estimated the single-year age-specific conditional probabilities of first marriage based on the assumption that the censored women (i.e. those who were still currently never married at the survey or census time) dropped out from the risk population at the middle of the year. (Elandt-Johnson and Johnson, 1980: 157).

9 Family Patterns

Guo Zhigang

China, with the largest population in the world, was a typical patriarchal familistic society for two thousand years. Throughout Chinese history, the family has played a crucial role in social and economic life and fulfilled the functions of production and consumption, organizing labour division and pooling individual risks, raising children and supporting the elderly.

It has long been assumed that in the past, extended family patterns were prevalent in China, and thus household size was in general substantially larger than today. The historical literature, including novels, dramas, state laws, and family records, overwhelmingly supports such an assumption. However, some other historical statistics seem to undermine this belief. According to the government registration of population and household records over the past two thousand years, national mean household size has ranged between three and seven, and been mostly around five (Liang Fangzhong, 1981; Zhao Wenlin and Xie Shujun, 1988). A study (Xia Wenxin, 1987), examining the historical statistics and selecting 71 of most accurate, found that most households had between five and six members. In the first half of the twentieth century, mean household size was around 5.3 (Ma Xia, 1984, 1988). These statistics show that very large extended families were not really the norm, and that although this family pattern may have been the social ideal it was only realized in the strata of nobles and wealthy people. High mortality and limited resources meant that ordinary households were smaller. It is increasingly recognized that the 'stem family' may have been the leading family pattern in China's history (Guo Zhigang et al., 1996), since it could also function to support the elderly in times when there was no formal support system.

Since 1949, great changes have taken place in China's economic and social system. Furthermore, the vast number of economic reforms and the open-door policy carried out since the late 1970s have resulted in changes in China's family statistics.

Definitions and Sources of Data

The family household is used as an instrumental definition for the family in this chapter because Chinese censuses, routine household registration and most surveys take the household as the basic enumeration unit. Most of the demographic data in this chapter is from China's population censuses and registration statistics. Descriptions of generational wealth flow are based on the survey of China's support systems for the elderly conducted by the China Research Centre on Ageing.

In China, households are classified into two categories: the family household is defined as a kin-based co-operative unit, and the collective household refers to institutional quarters such as dormitories in college, army, prison, and so forth. Occasionally, all-household data are used instead when family household data are unavailable. The interchange can be justified by the fact that over 99 per cent of households are family households, as shown in the censuses. So, the trends in overall household statistics approximate well to the trends of family households.

Shrinking Family Household Size

The 1953 national population census showed a mean household size of 4.3 in 1953, a drop of one person per household from the 1947 statistic of 5.3. It is believed that the rapid shrinking of mean size was due to the economic and political revolution (Guo Zhigang, 1995). In the six-year interval, the population increased at an annual average increase rate of 3.7 per cent, but the number of households increased faster, at an average rate of 7.6 per cent. The increment in household numbers during these six years was 47.6 million, equivalent to 55 per cent of the total number of households in 1947 (86.2 million). The major cause of the increase in the number of households may be attributed to the vast land reform in rural areas in the period 1950–2. At that time, over 90 per cent of national population were peasants. Most of the poor peasants (about 300 million) obtained redistributed land, housing, and equipment, which enabled them to build up new households.

In subsequent years the mean household size generally kept increasing until 1973, except for a period in the early 1960s; thereafter it declined (see figure 9.1).

The sharp decrease in mean household size in the early 1960s was due to demographic change. The failure of the Great Leap Forward, natural disasters and the breakdown of the relationship between China and the Soviet Union caused extreme economic hardship in these years. During this time, the birth rate dropped, death rates rose rapidly and the population increase rate became negative. The average number of people in each household became fewer.

On the other hand, it is worth noting that households increased in 1962 at an abnormally high rate (3.7 per cent). This peculiar change was almost

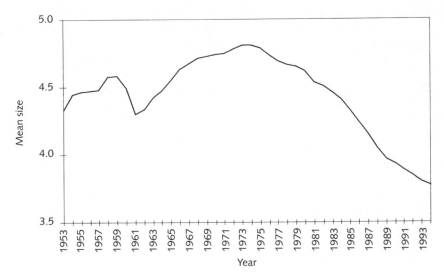

Figure 9.1   Mean household size, China, 1953–1994

certainly related to government policy in the face of severe food shortages. About 121,000 industrial enterprises, which had made up 38 per cent of the total in 1959, were closed. Urban residents were mobilized to move to rural areas to assist in agricultural production. As a result, the total number of workers was reduced by 18.9 million and total urban residents reduced by 26.0 million. This countryside-bound migration led many existing families to split and new households to form. Thus the number of households increased and the mean size of households decreased. However, while such policies stimulated family fission temporarily, they did not change the underlying mechanism of family formation. Therefore, when economic recovery ensued, this abnormal family fission stopped, split families gradually became reunified, fertility rose and mortality dropped. These adjustments led to the mean household size returning to the track of an increasing trend. This trend continued until 1973; then mean household size began a long-term decline.

The shrinking of family size can be explained in several ways. Some researchers attribute it mainly to the nuclearization of the family caused by social and economic changes (Pan Yunkang, 1984; Ma Xia, 1988). Some put more emphasis upon demographic change, namely, rapid fertility decline (Zeng Yi, 1986; Guo Zhigang, 1987). Some also take into account the interaction between the stem-family pattern and the changing population structure (Guo Zhigang, 1987; Guo Zhigang et al., 1996).

Demographic changes are evidently important factors in explaining the decreasing mean size of households, as household size is related to the dynamics of fertility. In 1971 total fertility started its steady decline, which was followed by a steady decrease in household size beginning in 1974. As children usually have to live with their parents as dependants, not doing anything to affect the numbers in a household, they can be regarded as an inactive population for purposes of household formation. Declining fertility reduced household size mainly by changing the proportion of children in the population.

Population census data show that the proportion of children (aged 0–14) decreased from 33.6 per cent in 1982 to 27.7 per cent in 1990. Correspondingly, the average number of children per household decreased from 1.49 in 1982 to 1.12 in 1990. Comparing the mean household sizes of 4.43 in 1982 and 4.05 in 1990, the average number of persons per household decreased by 0.38. But the reduction in the number of children on its own account for a decrease of 0.37. So reduced births during the eight years between censuses make up 97 per cent of the reduced mean household size. Therefore, fertility decline can be regarded as the major cause of the shrinking size of households or families. The provincial pattern of the family household size also relects the differentials in fertility level (figure 9.2).

Household Formation Level on the Basis of Headship

The adult population is the active population for household formation. Household headship among adults can reveal much about the overall level of household formation. The crude adult headship rates for China show an increase of 0.007 from 0.3452 in 1982 to 0.3524 in 1990, indicating that seven more households were formed per thousand adults in 1990 than in 1982. However, the headship rates by sex and age (see table 9.1 and figure 9.3) depict the different tendency towards joint rather than separate residence.

China's headship patterns have some obvious differences compared to Europe and North America (as represented by Sweden, Finland and the United States). First, the rates for both males and females are lower at young adult ages, reflecting the fact that young Chinese leaving their parental homes later with a sizeable portion never splitting off from them. Second, China's male rates show a distinct drop with age, which is attributable to the multi-generation family prominent in China. Many elderly people transfer head status to their adult children who live in the same household. Third, China's female rates are generally much lower. This phenomenon may be the result of many factors, such as: Chinese females' high dependency on their parents, their husbands, and their adult children; earlier marriage; the higher proportion of married women; the lower divorce rate, and so on.

The general pattern of China's specific rates has not changed very much, as can be seen by comparing the curves of 1982 and 1990. The rising rates for young males may be related to social changes such as migration to seek economic opportunity or higher education. The most remarkable decline took place in female rates at age 25 to 44. A few studies (Guo Zhigang,

Table 9.1 Headship rates by age and sex

Age	1982		1990	
	Male	Female	Male	Female
15–19	0.0225	0.0053	0.0195	0.0046
20–4	0.1796	0.0259	0.2403	0.0239
25–9	0.5688	0.0883	0.6366	0.0639
30–4	0.7940	0.1557	0.7944	0.0925
35–9	0.8712	0.1800	0.8589	0.1159
40–4	0.9011	0.1868	0.8902	0.1521
45–9	0.9146	0.1795	0.9032	0.1699
50–4	0.9031	0.1769	0.8925	0.1843
55–9	0.8821	0.1836	0.8610	0.1877
60–4	0.8237	0.1915	0.8182	0.1943
65–9	0.7433	0.2101	0.7459	0.1958
70–4	0.6558	0.2286	0.6408	0.2043
75–9	0.5649	0.2411	0.5622	0.2221
80–4	0.4877	0.2508	0.4803	0.2072
85+	0.4034	0.2550	0.3915	0.2262

1995; Goldstein et al., 1996) attribute the decline in female headship to the three waves of home-return migration brought about by government policies from 1982 to 1987. In the first, young people sent away to the countryside during the Cultural Revolution were helped to return home to cities. In the second, couples previously forced to live apart by job assignment were helped to reunite. In the third, one million persons returned to their home communities during the 1984–86 army demobilization. Statistics show that the percentage of married female heads of household who were living apart from their spouses was fairly high in 1982: 44.7 per cent in cities and 84.4 per cent in towns and counties. Yet in 1987 the corresponding figures were dramatically reduced to 21.1 per cent in cities and 63.2 per cent in towns and counties. Similar reductions occurred as well for the married male heads but with much smaller in original degree and decrements, 10.5 per cent to 3.1 per cent in cities and 4.1 per cent to 1.2 per cent in towns/counties. These figures indicate that in China women are more likely to become household heads when they are living apart from their spouse. When spouses are reunited, wives are more inclined to give up the head status.

There are only slight changes between the headship curves of 1990 and those of 1987 (1987 curves not shown). The percentages of married heads living apart from their spouses rose slightly in 1990 from the 1987 figure.

In general, household formation level on the basis of headship does not show the clear pattern change. More in-depth analysis shows that the

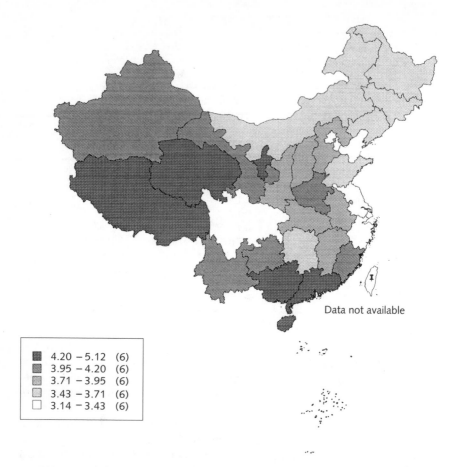

4.20 − 5.12 (6)
3.95 − 4.20 (6)
3.71 − 3.95 (6)
3.43 − 3.71 (6)
3.14 − 3.43 (6)

Data not available

Figure 9.2 Mean household size by regions, China, 1995

standardized adult headship rate of 1990 (0.3447) was lower than that of 1982 (0.3494), controlling for adult age and sex composition (Guo Zhigang, 1995). Similar results also come from standardized comparisons controlling for adult age, sex and marital status (0.3431 versus 0.3521). These results indicate the rise in crude adult headship rate from 1982 to 1990 was simply due to the compositional effect when the baby-boom cohorts entered the phase of high propensity to head their own households.

Living Arrangements and Family Pattern

Household formation and family pattern can be indicated by population distribution by family household type. Households are classified into

Table 9.2 Population distribution by age and household type, China, 1990 (%)

Age	Single	Couple	1-Gen	2-Gen	Q3-Gen	3-Gen	Sum
0–4	0.06	0.00	0.22	67.36	0.40	31.97	100
5–10	0.08	0.00	0.26	75.80	0.58	23.27	100
10–14	0.10	0.00	0.47	77.95	0.53	20.95	100
15–19	0.36	0.31	0.89	75.07	0.40	22.97	100
20–4	1.21	4.22	1.18	64.80	0.34	28.25	100
25–9	1.17	3.30	0.70	67.38	0.17	27.28	100
30–4	1.24	1.35	0.48	74.40	0.09	22.43	100
35–9	1.32	0.99	0.47	77.35	0.05	19.82	100
40–4	1.51	1.17	0.41	76.45	0.04	20.41	100
45–9	2.10	2.66	0.46	70.73	0.08	23.98	100
50–4	2.51	6.26	0.50	62.39	0.47	27.87	100
55–9	3.37	11.31	0.60	52.24	1.15	31.33	100
60–4	4.96	16.99	0.51	37.76	1.78	37.99	100
65–9	7.30	20.36	0.62	22.16	2.29	47.27	100
70–4	9.86	18.26	0.64	13.09	2.38	55.76	100
75–9	12.76	13.70	0.68	10.61	2.70	59.56	100
80–4	13.35	8.79	0.74	11.28	2.32	63.52	100
85+	16.28	4.40	0.73	14.91	2.78	60.91	100
Sum	1.60	3.29	0.58	66.98	0.50	27.04	100

several categories: (1) single person; (2) one couple only; (3) two-generation; (4) pseudo-three-generation (grandparents and grandchildren without the middle generation); (5) three-or-more-generation; (6) mixed one-generation (two or more individuals of the same generation such as siblings, or a combination of relatives and non-relatives).

In a one per cent sample of the 1990 population, only those living in family households were selected to aggregate the population distribution by household type within age groups, taking the age-group total as 100 per cent. This gives a more comprehensive picture showing the relative importance of each household type through the course of people's lives (see table 9.2 and figure 9.4). Because the percentages of pseudo-three-generation type and mixed one-generation type are very small, these two types are not discussed and not shown in the figure.

The percentage of single-person households increases as age rises and only attains any importance in the older phase. The curve of one-couple households passes through two peaks. The first one is much flatter showing a temporary expansion as people get married. After the first birth, young couples transfer to two-generation households. However, the second peak at old age represents a higher level and lasts a longer time. It results from children leaving the parental home. Finally, the increasing risk of widowhood turns down the percentage curve in very old age.

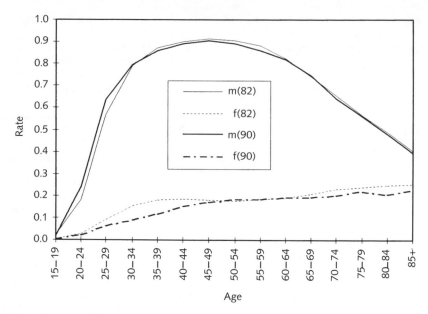

Figure 9.3 Headship rates by age and sex, China, 1982 and 1990

If we add the percentages of single-person and one-couple households together, we may obtain a useful indicator measuring the prevalence of 'empty-nest' families among older people. At the same time, this indicator also reflects the family nuclearization level in society as whole, because the empty-nest phase is a unique feature of the nuclear family life cycle, which does not appear in the life cycles of extended families and stem families. In addition, this indicator is not affected by population age composition. Among the very old, the percentage of one-couple households declines while the percentage of single-person ones increases, so that their number is relatively stable among the elderly aged 65 and above. This overall percentage for the old was 26.86 in 1990.

Among the age groups below 55, the people in two-generation households make up the majority in each age group. It should be noted that over 10 per cent of the elderly also live in two-generation households. This household type thus not only has the function of raising children but also supports the elderly. Besides, the two-generation household is not produced only by the nuclear family pattern, it can be a by-product of the stem family pattern as well. When the old have many children, only one of them may stay, while the others leave to build new households. This is the so-called interaction between population composition and stem family pattern, which may also result in increasing numbers of nuclear families. China is in just such a situation. Therefore, although two-generation households make up the majority, we cannot conclude that the nuclear family is the leading family pattern.

The percentage of three-or-more-generation households remains between 20 and 30 among the young and the middle-aged groups. For elderly people, however, this category represents the predominant form of living arrangement.

In summary, the majority of the old people live with their children. In the population aged 65 and above, those who live in households of made up of two generations or three or more generations made up 72.5 per cent of the total in 1990. The corresponding figure in 1982 was 73.1 per cent. On the other hand, the percentage of those living in 'empty-nest' families was 25.6 in 1982, but rose to 26.9 in 1990. It is clear that the nuclear family system occupies an important place in China and is gradually increasing. However, the majority of the Chinese elderly still live with their adult children. This phenomenon reflects the tremendous influence of a deeply rooted culture and the stability of the traditional family as the leading pattern in society.

Children Alive as Determinants of the Living Arrangements of the Elderly

Demographic availability is one constraint on residential arrangements for the elderly (Kobrin and Goldscheider, 1982). An analysis based on a one per thousand sample of 1990 confirms that both the number and sex of children affect living arrangements. This study was, however, confined only to women aged 60 to 64, as the census information was unavailable for other age and sex groups of the elderly (Guo Zhigang, 1996a). The majority of these women live in one of the four major household types discussed in the previous section. Figure 9.5 shows the relation between the number of children and household types grouped into two categories: empty-nest and co-resident with children.

The percentages of household type are strongly correlated with number of offspring. Among childless women, about half are living in empty-nest households, indicating that some old people live involuntarily without children. However, in cases where there is one child available, the percentage drops by almost a half. When there are more children and an optional choice, the elderly show an even greater propensity to co-live with one or other of them. Nevertheless, 20 per cent of the elderly who have three or more children live apart from them. This figure may be taken as a better indicator for the prevalence of the nuclear family pattern.

An inconsistency in the statistics warrants special attention. A significant number (39 per cent) of childless women in fact live with children. It is unlikely that an elderly woman would not report her own children as alive while they were living together with her in the same household. Careful investigation reveals that 90 per cent of this group also reported never having any children ever born. This eliminates the possibility that these

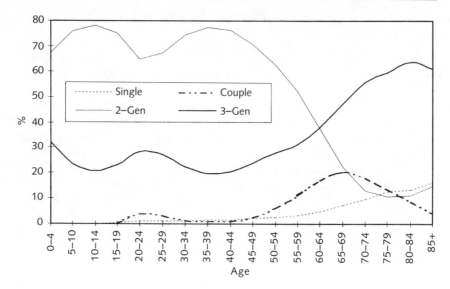

Figure 9.4 Population percents by household type within age groups, China, 1990

elderly women were living with step-children after their own had children died, since a step relationship is also defined in the census as one of parent and child. Therefore it seems likely that these co-living children are adopted. This further shows the importance of the support the elderly receive from children.

China has long been patriarchal society with a strong son preference. The sex of children also influences the living arrangements of the elderly. For example, the study cited above found that of older women who have at least one son alive but do not have a daughter, 21.4 per cent live in empty-nest households and 75.6 per cent live in co-resident households. The corresponding figures are 32.4 and 60.1 for those who have at least one daughter alive, but do not have a son. Therefore it is more common for the elderly to live with sons than with daughters. This shows that son preference is not merely confined to ideology, but is also rooted in people's lived experiences. However, such a situation has been changing gradually for several decades, especially in cities. At present, quite a large proportion of those who have daughters only live in multigenerational households, with their daughters and sons-in-law taking the responsibility for elderly support. According to a retrospective survey in five big Chinese cities since 1958 (Liu Ying, 1987), the proportion of young couples living with the brides' parents has been increasing. The proportion of those who got married between 1958 and 1965 is 12.44 per cent; the corresponding proportions for the 1966–76 and the 1977–82 marriage cohorts are 14.14 per cent and 18.23 per cent, respectively. In addition, some young couples invite the wife's parents (mostly the mothers) to move in to take care of the children for a period after marriage.

Inter-Generational Wealth Exchange during Old Age

Data from the 1992 old people survey by the China National Ageing Study Centre have been used to compute economic exchange between generations. The survey defines the elderly as people aged 60 and above, and draws on 9,889 cases from cities and 10,194 cases from rural areas. A few studies tend to stress the fading importance of economic support from adult children to the elderly based on the survey data (Xia Chuanlin and Ma Fengli, 1995). However, this study simply took the elderly as a homogeneous group. While taking age into account, the wealth exchange patterns show obvious variation by age (Guo Zhigang, 1996b). Net-flow is defined as the differential of gross up-flow minus gross down-flow measured in equivalent monetary value. Three classifications are made on the basis of value of net-flow: down-flow when the net-flow is negative; up-flow when it is positive; and even-flow when the net equal 0, which indicates two kinds of cases: (1) no flow at all; (2) the up- and down-flows are even (see table 9.3).

In the city samples, the older people are, for the most part, obtaining a net gain from their children. The even-flow category encompasses two sorts of situation. In the one instance, it turns out to be a transitive phase, because more and more people are transferring from the down-flow class to the up-flow class with ageing via this middle phase. In the other, it shows a relatively stable portion of the elderly who remain in the same situation throughout later life. As for the down-flow class, the proportion decreases steadily by age. The age-specific patterns of inter-generational wealth flow advance our knowledge about the changing conditions of the elderly.

Table 9.3 Percentage of the elderly by net–flow type, China, 1992

Age	City				Village			
	Down	Even	Up	Sum	Down	Even	Up	Sum
60–4	34.5	37.7	27.8	100	9.4	27.8	62.8	100
65–9	24.0	37.5	38.5	100	7.1	22.5	70.4	100
70–4	17.4	37.5	45.0	100	4.7	19.9	75.4	100
75–9	12.2	32.3	55.5	100	2.5	13.9	83.5	100
80–4	7.6	31.0	61.4	100	2.4	13.5	84.1	100
85+	6.0	23.1	70.9	100	0.8	10.8	88.4	100
Overall	**23.9**	**36.4**	**39.7**	**100**	**6.3**	**21.8**	**71.9**	**100**

*Those who do not have surviving children are excluded.

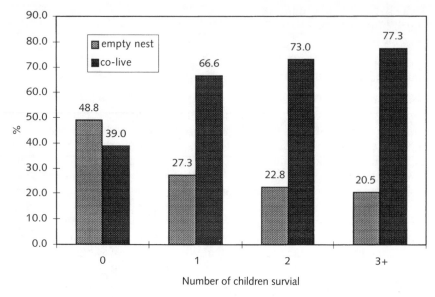

Figure 9.5 Family type percent of women aged 60–64 by children number, China, 1990

Without control by age, the overall proportions tend to over-stress the situation of the younger old, because there is a stronger compositional effect within the elderly population. In the sample, 75 per cent of cases in total are aged 60 to 64; they are not old enough to have experienced the difficulties connected with ageing.

The rural sample shows that the rural elderly rely more upon their children because there is less down- and even-flow, and more up-flow. The difference in exchange pattern between city and rural areas reflects the difference in economic development and social security of the elderly.

These findings are not only relevant to show the situation in China, but also pose methodological questions for ageing studies in general.

Summary and Discussion

Currently in China, mean household size is shrinking mainly due to fertility decline and compositional changes in the population. The stem family remains the leading pattern in society and most of the elderly still live with their adult children. The nuclear family has taken an important place in society, and this pattern is gradually expanding. The presence of children is an important factor in maintaining the stem-family pattern, so the future family pattern will be strongly affected by the rapid change in population.

Up to the present, the traditional method of elderly support by upward wealth flow from adult children is still dominant in society, especially in rural areas. This fact should be given attention in policy-making relating to the family and social security. This chapter only describes the economic relations among generations; providing support for the elderly in the shape of daily care and emotional comfort is also an important function of the family.

Facing rapid population ageing, it is advantageous for China to integrate social development with family development. On one hand, the social security system for all of the elderly should be developed and improved. On the other hand, it is important to maintain multigenerational families and facilitate families' own ability to support their elderly relatives by favourable policies and public aid.

References

Goldstein, Alice, Guo, Zhigang and Goldstein, Sidney (1996), *The Relation of Migration to Changing Household Headship Patterns in China, 1982–1987*. Population Studies and Training Centre of Brown University working paper series: no. 96-01.

Guo, Zhigang (1987), Several problems in the study of family households, *Population Research*, 2, 10–14 (in Chinese).

Guo, Zhigang (1995), *Contemporary Population Development and Family Household Change in China*, Beijing: Renmin University Press (in Chinese).

Guo, Zhigang (1996a), Factors affecting the living arrangement of elderly Chinese women, *Population Research*, No. 5 (in Chinese).

Guo, Zhigang (1996b), Analysis of inter-generational wealth flows during old age, *China Gerontology Journal*, No. 5 (in Chinese).

Guo, Zhigang, Goldstein, Alice and Goldstein, Sidney (1996), Changing family and household structure. In Alice Goldstein and Wang Feng (eds), *China: The Many Facets of Demographic Change*, Boulder, CO: Westview Press.

Kobrin, F. E. and Goldscheider, C. (1982), Family extension or nonfamily living: Life cycle economic and ethnic factors, *Western Sociological Review*, 13 (1), 103–18.

Liang, Fangzhong (1980), *Statistics of Household, Farmland and Land Tax in China's History* (in Chinese). Shanghai: People's Publication House.

Liu, Ying (1987), Marriage and family in China's cities. In Population Research Centre, CASS (ed.), *Almanac of China's Population*, Beijing: Social Science Literature Press (in Chinese).

Ma, Xia (1984), An analysis of the size of family households and family structure in China, *Population Research*, 3, 46–53 (in Chinese).

Ma, Xia (1988), The development of family size and structure. In Xu Dixin (ed.), *The Population of Contemporary China*, Beijing: China Social Science Publication House (in Chinese).

Pan, Yunkang (1987), Popularizing trend of nuclear family in China cities. In Liu Ying (ed.), *Studies of Marriage and Family in China*, Beijing: Social Science Literature Press, 129–42 (in Chinese).

Xia, Wenxin (1987), Preliminary study of family size. In Liu Ying (ed.), *Studies of Marriage and Family in China*, Beijing: Social Science Literature Press, 100–12 (in Chinese).

Zeng Yi (1986), Changes in family structure in China: A simulation study, *Population and Development Review*, 12 (4), 675–703.

Zhao, Wenlin and Xie, Shujun (1988), *China's Population History* (in Chinese). Shanghai: People's Publication House.

10   Education in China

Peng Xizhe

General Situation

All Chinese governments since the 1920s have proclaimed their commitment to universal elementary education, but it is only in the post-1949 era that the commitment has been basically realized (Pepper, 1990). Great efforts have been made to expand basic education, and remarkable success has been achieved in raising the educational level of the Chinese people during the past half a century. The improvement has been in general continuous apart from interruptions that occurred during the 1959–61 famine and the Cultural Revolution periods. In 1949 the total number of students enrolled in all educational institutions was 2.58 million. The number (excluding those in adult education) has increased to the point where there were more than 200 million students studying in 0.8 million educational institutions by the end of 1995. These figures make China's education system one of the largest in the world.

The illiteracy rate for people aged 15 years and over had declined to 22.21 per cent in 1990, and has further decreased year by year since then. In 1990 the total number of illiterate people aged 15 to 47 in China was around 35 million, and the number has declined by 5 million annually since the early 1990s. On the whole, illiteracy is a problem overwhelmingly among the older age groups. 90 per cent of illiterate people are peasants, and concentrated mainly in a few western provinces.

China's success in the improvement of universal literacy has been achieved mainly through a strong commitment to the widespread provision of elementary education at an early stage of development. Compared to India, for example, China is well ahead in the field of basic education (Dreze, 1995). In 1990 almost half of the adult population in India was illiterate, compared with only 22 per cent in China. Chinese government policy with respect to education at the various levels is: to vigorously strengthen basic education; to actively develop vocational and adult education; to ensure appropriate development of senior secondary schools; and to improve institutions of higher learning with the stress on intensive development.

The emphasis in China's education system has changed over time. In the 1950s the main aim was eliminating illiteracy. It later changed to compulsory primary education, and a further shift to universal nine-year education took place after the mid-1980s. Since early 1999, a new strategy focusing on 'education in qualities' has been widely advocated by China's education authority, and is also supported strongly by the top Chinese leadership.[1] A special national Work Conference on Education was convened to push forward the implementation of this action plan.[2] The new approach emphasizes the importance of originality and creativeness, rather than a knowledge-centred or examination-oriented education. It is the response of the Chinese government to the development of a knowledge economy that is characterized by the rapid expansion of information technology and life sciences.

China's basic education experienced rapid expansion during the 1970s, but the quality of education deteriorated as a result of the general social disturbance caused by the Cultural Revolution. Since 1986 nine-year compulsory education has been regarded as a basic right of all Chinese citizens and a fundamental policy of the State, as was clearly indicated by the Law of Compulsory Education. The targets, set by China's Education Reform and Development Plan in 1992, are to eliminate the remaining 36 million adult illiterates by the year 2000 and to provide nine-year compulsory education to all school-age children across China by the early twenty-first century. By the end of 1995, it was reported that 1025 Chinese counties or regions, covering 36 per cent of the total population, had reached the target. Further efforts have been focused on poor rural areas. Equality of educational opportunity between boys and girls has been achieved for China as a whole, but the general enrolment rates for girls are somewhat lower in many areas.

It is reported that at present primary education is functioning well in areas containing 91 per cent of the total Chinese population, and the enrolment rate for primary school-age children[3] has reached as high as 98 per cent, compared to 49 per cent in 1952 (table 10.1). However, school attendance rates are 8 percentage points lower than the net enrolment rate, which indicates that there is a problem with school dropout (Tsui, 1997)

The Education Structure of China

China's current education system has four major components:

1. Elementary education consisting of pre-school, primary and middle-school education.
2. Technical and vocational education including vocational schools, technical schools and polytechnics. Various in-service training programmes also belong in this category.

Table 10.1 Enrolment of school-age children* (in 1000s), 1952–1996

Year	1952	1965	1980	1985	1990	1992	1996
School-age population	66,424	116,032	122,196	103,623	97,407	111,562	128,765
School enrolment	32,681	98,291	114,782	99,428	95,948	108,455	127,233
Enrolment (%)	49.2	84.7	93.9	96.0	97.8	97.2	98.8

*School-age children refers to 7–13 year olds in general, but varies over time and between provinces.
Source: Hu Ruiwen (1996), *The New Development in China's Education during the 1980s and prospect of the 1990s*. Shanghai Institute for Intelligence Development Research.

3. Higher education consisting of undergraduate and postgraduate studies.
4. Adult education including activities aimed at the elimination of illiteracy and various formal and informal training programmes for adults.

China adopted an American-style school system in 1922 and now has a 6–3–3–4 structure in general (Gao, 1985). Pre-school education is provided in nurseries and kindergartens which are in general supported or organized either by parents' work units or local communities. The length of schooling at most primary schools is 6 years. Primary school entry is at age 6 (or 7 in some places). General secondary school is divided into two stages, 3 years at junior middle and 3 years at senior middle school. University education is normally for 4 years, though there are some exceptions. The duration of the academic year is fixed across China. Usually it runs from September to June/July, and is divided into two semesters.

Examinations are carried out at each educational level and have played a very important role in determining the fate of students – except during the Cultural Revolution period when all examinations were abolished. More recently, various experimental measures have been taken, mainly in primary schools, to reform the examination system in order to reduce the burden on young students while at the same time improving the general quality of education at each level. The general responses to these changes have differed.

The vast size of the population has made it difficult to achieve consistency and coherence in educational matters. Generally, the education system of China has provided for linguistic diversity. In minority regions, minority national languages are used in the classroom while standard Mandarin is used as the second language. The textbooks used in minority regions are specially written under the general guidance of the curriculum.

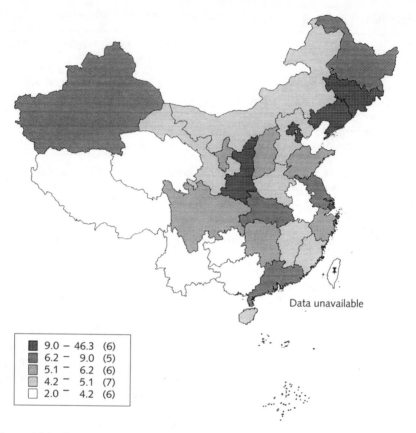

Figure 10.1   Percentage enrolment for people aged 18–21, 1994

Vocational schools (secondary specialized schools) and upper secondary schools provide different courses for children over the age of compulsory schooling. While students in upper secondary schools are usually aiming at a university education in the future, students in vocational schools are trained as professional workers and other middle-level staff. Since the scale of China's higher education is not big enough to absorb all high-school graduates, there has been a shift in priorities from general education to specific technical preparation at particular levels. Secondary specialized schools increased in number from 1710 in 1952 to 4049 in 1995, and total student enrolment rose from 636,000 to 3.72 million over the same time period. Vocational higher education is to be given priority in the coming decade to produce much-needed trained personnel in China's economic sectors.

Table 10.2 Basic statistics of regular schools by level and type (in 10,000s, except for number of schools), China

	No. of schools	No. of graduates	Entrants	Total enrolment	Teachers, staff and workers	
					Total	*Of which: full-time teachers*
Total	**911,174**	**4,050.53**	**6,972.70**	**24,072.30**	**1,455.81**	**1,141.37**
I. Graduate education	**735**	**4.65**	**6.37**	**17.64**	—	—
1. Institutions of higher education	412	4.32	5.92	16.33	—	—
2. Research organizations	323	0.33	0.45	1.31	—	—
II. Regular institutions of higher education undergraduates	**1,020**	**82.91**	**100.04**	**317.38**	**103.15**	**40.45**
III. General secondary schools	**97,308**	**2,000.03**	**2,575.27**	**7,188.26**	**587.14**	**430.12**
1. Specialized sec. schools	4,143	115.71	162.11	465.41	55.36	27.64
Secondary technical schools	3,251	86.32	129.57	374.32	43.96	21.25
Teacher training schools	892	29.38	32.54	91.09	11.40	6.39
2. Skilled worker schools	4,395	69.94	73.40	193.10	30.95	11.57
3. General sec. schools	78,642	1,664.04	2,128.20	6,017.86	454.09	358.67
Senior	13,880	221.66	322.61	850.07	—	60.51
Junior	64,762	1,442.38	1,805.59	5,167.79	—	298.16
4. Vocational schools	10,047	150.10	211.22	511.89	46.74	32.24
Senior	8,578	129.17	180.34	431.00	—	28.23
Junior	1,469	20.93	30.88	80.89	—	4.01
5. Correctional work-study schools	81	0.24	0.34	0.63	0.27	0.15
IV. Primary schools	**628,840**	**1,960.14**	**2,462.04**	**13,995.37**	**643.63**	**579.36**
V. Special education schools	**1,440**	**2.80**	**4.61**	**34.06**	**4.33**	**2.85**
VI. Kindergartens	**182,485**		**1,824.37**	**2,518.96**	**117.29**	**88.44**

Source: China Education Statistics Yearbook, 1997, China State Education Commission, also see http://www.moe.edu.cn.

In addition to this regular education system, China has a large-scale and well-developed adult education system. It offers education to the adult population ranging from literacy classes to higher education. Table 10.3 presents the basic statistics for adult education in China in 1997. It can be seen from the table that about 70 million Chinese adults are taking part in various educational activities. It is projected that more adults will participate in higher education through this system in the coming years. In addition, education for the elderly population is also a flourishing business as China enters the ageing society.

The Management Structure of China's Education System

China's educational administration, particularly in higher education, used to be highly centralized. Educational authorities were established in parallel with the government's administrative hierarchy. From the centre down to the township, education bureaux, however they are labelled, are consistently one of the major divisions of government at each level. The State Education Commission, which was renamed the Ministry of Education in 1998, is the ministry under the State Council responsible for the overall administration of education affairs in China. In addition to the network of education bureaux, education departments also exist under various other ministries in order to manage the vocational and technical schools, as well as some specialized institutions of higher learning that come under the direct control of these ministries. With the flourishing of adult education, these education departments are playing a more important role than ever before. This situation is undergoing changes resulting from the general restructuring of China's government organization. Nowadays, the Ministry of Education has taken over responsibility for all educational matters, including the guidance for those institutes of higher learning that used to belong to other government ministries.

In accordance with the general trend of decentralization in the era of China's economic reform, local governments and universities themselves have taken more responsibility for the management and financing of education. The principal function of the Education Commission (Ministry of Education since 1998) is to plan and co-ordinate educational development at each level, rather than to run schools directly. However, schools at a given level use the same textbooks and follow the same curriculum, determined most likely by the provincial education authority. Students' educational achievements are likewise evaluated using the same criteria across administrative boundaries.

In addition to the educational institutions run by the government at each level, many other forms of schools have emerged with the progress of China's reform, covering every level of China's education system and ranging from kindergartens to universities. These schools are under the general guidance of the education authorities at each level, but their

Table 10.3 Basic statistics of adult schools by level and type (in 10,000s, except for number of schools), China

	No. of schools	No. of graduates	Entrants	Total enrolment	Teachers, staff and workers	
					Total	Of which: full-time teachers
Total	**657,751**	**9,541.66**	**8,024.29**	**7,008.1**	**117.92**	**48.92**
I. Higher educational institutions for adults	**1,107**	**89.20**	**100.36**	**272.45**	**21.46**	**10.02**
1. Radio/TV universities (RTVU)	45	18.46	19.36	50.60	4.97	2.35
2. Workers' colleges	664	11.46	12.00	32.55	8.39	4.02
3. Peasants' colleges	4	0.04	0.04	0.09	0.03	0.01
4. Institutes for administration	161	6.04	7.39	16.07	3.60	1.40
5. Educational colleges	229	7.42	8.65	22.83	4.33	2.17
6. Independent correspondence colleges	4	0.43	0.46	1.37	0.14	0.07
7. Run by regular institutions for higher learning	—	45.35	52.46	148.95	—	—
Divisions of correspondence	—	27.30	30.45	90.32	—	—
Evening schools	—	8.86	11.49	34.45	—	—
Short-cycle courses for adults	—	9.19	10.52	24.18	—	—
Of the total: regular full-time short-cycle courses provided by RTVUs for sec. school graduates	—	7.00	7.79	19.10	—	—
II. Secondary education for adults	**462,993**	**8,779.99**	**7,323.43**	**6,119.59**	**75.36**	**32.47**
1. Specialized sec. schools for adults	5,113	91.22	104.92	266.38	23.55	12.68
Radio/TV specialized sec. schools	152	18.13	28.29	70.24	1.90	0.89
Specialized sec. schools for staff & workers	2,026	31.40	36.67	92.07	8.97	4.63
Specialized sec. schools for cadres	263	6.13	6.37	15.25	1.42	0.68
Specialized sec. schools for peasants	457	7.77	9.43	22.49	1.97	1.23
Correspondence specialized sec. schools	73	7.55	7.72	21.97	1.29	0.54
In-service teacher-training schools	2,142	20.24	16.44	44.36	8.00	4.71

Table 10.3 continued

	No. of schools	No. of graduates	Entrants	Total enrolment	Teachers, staff and workers	
					Total	Of which: full-time teachers
2. General secondary schools for adults	5,866	109.51	95.57	126.33	5.52	2.21
General sec. schools for staff & workers	1,922	64.09	43.52	73.13	4.07	1.33
General sec. schools for peasants	3,944	45.42	52.05	53.20	1.45	0.88
3. Technical schools for adults	452,014	8,579.26	7,122.94	5,726.88	46.29	17.58
General sec. schools for staff & workers	10,858	557.88	608.30	386.22	8.12	4.33
Technical training schools for peasants	441,156	8,021.38	6,514.64	5,340.66	38.17	13.25
III. Adult primary schools	**193,651**	**672.47**	**600.50**	**616.06**	**21.10**	**6.43**
1. Worker primary schools	1,087	11.01	14.76	15.13	0.36	0.19
2. Peasant primary schools	192,564	661.46	585.74	600.93	20.74	6.24
Of which: Literacy classes	140,761	403.54	314.77	353.00	15.10	4.18

Sources: China Education Statistics Yearbook, 1997, China State Education Commission; also see http://www.moe.edu.cn.

governing bodies determine the recruitment of teachers, financial management and curriculum development. The ownership of these schools varies
from private, to partnership and collective. In 1996 there were 60,000 privately owned education institutes offering education to 9.8 million students,
accounting for 0.6, 1.2, and 2.8 per cent of total students in junior-middle,
high-middle, and vocational schools, respectively. In addition, there are
increasing numbers of private educational agencies offering various kinds
of training, and personal tuition etc. It seems very likely that this trend will
continue in the near future.

The Development of China's Higher Education

China's entire education system was in chaos during the Cultural Revolution period (1966–76), but higher education suffered the most. The entry
examination to universities was abolished and students were evaluated not
by their academic performance but according to their political activities.
General order was restored after 1976, and nationwide higher-education
entrance examinations were reintroduced in 1977 in order to select the best
students.

After 1977 the number of higher education institutions steadily increased
from 404 to 1063 in 1993, while enrolment at undergraduate level rose from
625,000 to 2.066 million, and jumped further to 2.90 million in 1996 (see
table 10.4). Postgraduate students have also risen from 93,000 in 1990 to
162,000 in 1995, studying in 740 qualified institutions of higher learning.
Between 1979–93, over 6.22 million people graduated from colleges or
graduate schools, 1.1 times the total number for the period prior to reform
(*Beijing Review*, 1994).

Since the early 1950s, there has been a general tendency in China's higher
education system to over-emphasize the natural sciences while neglecting
the social sciences. The percentage of comprehensive universities offering
both natural and social sciences programmes declined from 23.9 in 1949, to
10.9 in 1952, and further down to 4.3 in 1986. Meanwhile, the proportion of
students majoring in social sciences in relation to the total number of university students stood at 33.1 per cent in 1949, 6.8 per cent in 1962 and 8.9
per cent in 1980 (Yang, 1997). A similar situation was also to be seen in the
curriculum structure of the middle, and even the primary schools. At the
same time, specialization was strengthened and generality weakened in
teaching arrangements and degree requirements. On the whole, this tendency has already shown its negative impact on China's modernization
process. Some counter-measures has been discussed and taken to cope with
the issue. The New Action Plan for Chinese education stresses the need to
improve Chinese higher education by training students to have initiative,
broad knowledge, social responsibilities, and adaptability.

The mid-1990s witnessed a new wave of restructuring and decentraliz-
ing in China's higher education. Local governments at provincial level have

Table 10.4 Growth of China's higher education, 1952–1996

		1952	1965	1980	1985	1990	1996
Number of higher education institutions		201	1,192	3,357	2,232	2,396	2,170
Conventional institutions		201	434	675	1,016	1,075	1,032
Post-graduates* (1,000s)	Total	—	5	22	87	93	162
	PhD	—	—	—	4	11	35
	Master	—	—	—	75	81	127
Under-graduates and	Bachelors	—	—	—	1,331	1,532	2,068
	Conventional	131	644	862	1,123	1,320	1,795
	Adult	—	—	—	208	212	273
diploma students* (1,000s)	Diploma	—	—	—	2,097	2,198	3,609
	Conventional	60	30	282	580	743	1,226
	Adult	—	—	—	1,517	1,455	1,316
	Total	191	1,087	1,641	3,428	3,730	5,667

*Number of enrolled students.
Sources: As table 10.1.

taken on far more responsibility for the managing and financing of universities located in their provinces. In return, universities are required to work much more closely with local governments rather than with the central education authority in Beijing, to serve for the needs of local socio-economic development. Attempts were also made to create new-style universities. For example, one aim was to make the universities comprehensive, in the sense of including many different disciplines. Mergers have been widely discussed and practised to promote quality and economies of scale in China's higher education, which has already resulted in the decline of the total number of high learning institutes. Consequently, the average number of students in each university has increased from 2000 in the early 1990s to 3070 in 1996. Some very large universities, such as new Sichuan University and the new Zhejiang University, are already in full operation.

Compared with its large population, the scale of China's higher education is rather small. In 1994 only 2.4 per cent of all young people aged 18–22 had the opportunity of being admitted into university education. This figure is even below the average value for developing countries. According to the new action programme for education development, the average number of students in each of China's universities should reach 4000, so that a total of 6.6 million students can be accommodated in the year 2000. This will increase the enrolment rate for higher education to 15 per cent by the same year.

Parallel to the regular system is a highly developed informal system of part-time institutions designed particularly for continuing education for

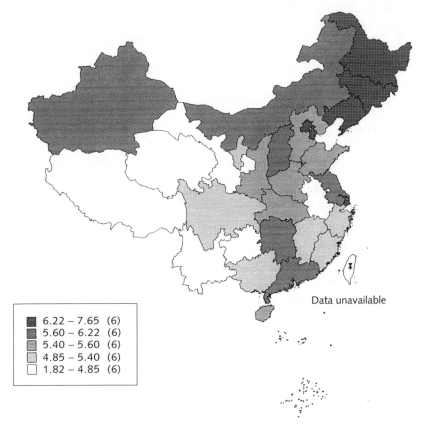

6.22 – 7.65 (6)
5.60 – 6.22 (6)
5.40 – 5.60 (6)
4.85 – 5.40 (6)
1.82 – 4.85 (6)

Data unavailable

Figure 10.2 Average number of years of schooling, China, 1990

adults. There is a special television network, exclusively for educational use, covering most parts of China. The curricula of these informal education institutions are similar to those in the formal education system, but students take longer to complete their education and be awarded diplomas or degrees. In 1997 there were a total of 1,107 institutions of higher learning dedicated to adult education accommodating 2.72 million adult students (see table 10.3).

China sent its first group of students to study abroad in 1872. This practice was interrupted entirely during the Cultural Revolution period. It was in 1978 that China re-opened its door to the outside world. In that same year, more than 800 Chinese were sent to study in over 20 countries including the USA, UK and a few other western countries. Since then, about 300,000 Chinese students and scholars have been sent abroad and have studied at various academic institutions in more than one hundred countries. So far, about one third of them have returned to China, and they have become one of the most influential groups in China's development.

At the same time Chinese universities have started to receive foreign students. In 1997 43,000 foreign students were admitted into 330 Chinese institutions of higher learning. It was reported that a total of 258,000 foreign students had been studying in China during the last 20 years. International co-operation in teaching and research matters also experienced rapid development. Many United Nations organizations have been conducting education projects in China. For example, the United Nations Population Fund has invested US$17.2 million to promote demographic training and research in Chinese Universities. The World Bank is the largest co-operative partner of the Chinese education sector. More than US$1.38 billion in loans from the Bank have been utilized since 1981 when the first World-Bank-supported education project began implementation (*Beijing Review*, 1999).

Education Finance

Lack of financial investment has always been a bottleneck in China's educational development. The total education expenditure for China as a whole, in absolute terms, has increased steadily during the 1990s, from 73.15 billion in 1991 to 187.8 billion in 1995, and further to 253.173 billion in 1997. In 1994 alone a 40.46 per cent annual increase was recorded (see table 10.5). However, the share of education expenditure in GDP reveals a different situation. The target set by China's Education Reform and Development Plan of 1992 is to make overall education expenditure account for 4 per cent of total GDP by the year 2000. The actual figure, however, has been around 2 per cent since the 1980s. It was 3 per cent in 1992, and dropped to 2.76, 2.52 and 2.46 for the following three years, respectively.

Table 10.5 Sources of education investment (in billion yuan), 1994–1995

	1994	*1995*
Government direct education budget	88.40	102.84
Education tax levied at various administrative levels	13.28	18.91
Other government resources	0.82	1.23
Education input from enterprises	8.91	10.49
Collections from education services	6.07	7.68
Donations from NGOs and individuals	1.08	2.04
Collections within communities	9.75	16.28
Tuition and other fees	14.69	20.12
Others	5.89	8.20
Total	148.88	187.80

Source: UNDP *China: Human Development Report, Human Development and Poverty Alleviation, 1998.*

Direct government expenditure has always been the most import financial resource for education. In 1995 the Chinese government spent 102.84 billion, or 16.05 per cent of its total budget, on the education sector, of which 12 per cent came from central government and 88 per cent from local government. There is a steady upward trend both in absolute and percentage terms in the long run. In 1997 government expenditure accounted for 73.56 per cent of the total education input. However, the increase in direct government investment in education has not been as fast as the increase in overall government expenditure, which was also a target set by the 1992 Plan. It is therefore a difficult task for the government to increase the share of education expenditure to 4 per cent of total GDP, as targeted, by the year 2000. The education authority announced recently that the ratio of education expenditure in the annual fiscal budget of the Central Government would increase one percentage point each year from 1998 to 2002. And local governments have been urged to increase their own funds on education (*People's Daily*, 1999).

In 1995 16.22 per cent of government input was directed to higher education, 12.12 per cent to middle-level vocational education, 9.37 to conventional middle schools, 5.78 to pre-school education and the largest part of the investment, 56.51 per cent, was used to support nine-year compulsory education.

Since the mid-1980s, the financing of basic education has increasingly been the responsibility of local government. Local governments, including the governance committees of rural villages, began to levy an education surcharge, which is linked to indirect taxation and became the major resource of rural primary education. As many rural communities failed to collect this surcharge from the poor peasant households or diverted the funds to other uses, this financing system enlarged the gap in education attainment between rich and poor areas. As a result, the maintenance and improvement of basic education in poor regions, especially poverty-stricken rural areas, are facing a serious challenge.

It has been reported that the central government will invest an additional 3.9 billion yuan in a special education fund to support the poorest areas and enable them to fulfil their education targets. Together with local input, the total government education investment in poor areas will reach more than 10 billion, which will be one of the largest direct government investments in China's elementary education in recent years.

In addition to direct input from the government, the whole of society has been mobilized to support China's ambitious plan for nine-year compulsory education and the elimination of illiteracy among young people, especially in the remote poor rural areas. The nationwide fund-raising programme named the 'Hope Project' is one successful example of collaborative efforts of this kind.

The 'Hope Project' was formally launched in October 1989, organized by the China Youth Development Foundation. It aims at collecting donations from individuals, families and social organizations to help children in China's poor rural areas, especially the non-attenders at schools, to complete their nine-year compulsory education. The project has achieved great success both in terms of raising funds and raising mass awareness and participation. According to the evaluation report,[4] fund-raising reached its peak in 1994, and most of the donors joining the project through one-to-one direct linkage with the beneficiaries. It is reported that, by the end of 1997, the project had received 1.257 billion RMB yuan in donations at home and from abroad, from institutions and tens of millions of individuals from all walks of the society. More than 70 per cent of the donors are urban residents or organizations. During the same time period, the project has provided 975 million yuan as financial support to 1.847 million students to enable them to return to school, and sponsored the construction of 5256 schools, which are named the Hope Primary Schools. Roughly 31 per cent of non-attenders in 695 rural counties were supported by the project in 1996, while the rate in 1990 was only 1 per cent. Meanwhile, scholarships have been given to nearly 10,000 high school and university students to help them complete their education. Even though there are shortcomings in the operation of the project, it does make a substantial contribution to China's educational development.

Other NGOs targeting special population groups set up similar funds. According to the 'Hope Project' report, girls accounted for more than half, 53.3 per cent, of the total number of young people not attending schools, but only 45.8 per cent of project-supported students were female. Given this background, the All China Women's Federation established a similar fund, the 'Spring Bud Project', to safeguard the educational rights of girls in poverty-stricken areas.

Previously in China, higher education was free to students. This meant that the State was responsible for all public and personal expenses related to higher education. In return, all graduates had to take the jobs assigned by the education authority in accordance with the general central planning system. Reform began in 1994 when students in 38 universities were asked to pay a certain amount of tuition fees themselves. The number of universities joining this reform increased to 500, and by 1997 all university students were expected to pay part of their tuition fees. Meanwhile, a comprehensive scholarship programme, and a study-loan system as well, have been developed to provide basic financial assistance to those in need and those with good academic achievement. Graduates, meanwhile, are given the freedom to look for the jobs in the labour market.

Diversities in Educational Achievement

Due mainly to the differences in educational infrastructure and historical background, which run parallel with general socio-economic development, there are marked discrepancies in regional educational achievements (see table 10.6). The interprovincial disparity has increased since the beginning of the economic reform, due mainly to changes in the financing system for

Table 10.6 Regional difference in education

	Literacy rate[1] (%)	Average years of schooling[2]	Enrolment for children aged 8[3]	Enrolment rate in higher education[4]	Graduation rate for primary school[5]
Beijing	92.1	7.65	99.1	46.3	100
Tianjin	90.7	6.91	99.5	20.64	99
Hebei	86.7	5.50	96.7	4.92	98
Shanxi	90.6	6.04	93.3	5.52	93
Inner Mongolia	83.4	5.67	91.3	4.53	91
Liaoning	90.7	6.47	95.3	11.35	97
Jilin	90.7	6.26	92.6	10.18	98
Heilongjiang	89.2	6.22	92.7	8.61	94
Shanghai	91.5	7.27	100	23.75	100
Jiangsu	85.4	5.60	98.6	6.61	95
Zhejiang	83.0	5.30	96.7	5.61	94
Anhui	80.6	4.57	87.9	4.03	89
Fujian	79.4	5.15	93.7	5.09	86
Jiangxi	81.4	5.14	88.2	4.73	79
Shandong	82.4	5.40	94.3	6.17	94
Henan	84.1	5.48	97.0	4.20	86
Hubei	82.9	5.58	85.0	8.90	93
Hunan	84.7	5.66	95.5	6.09	92
Guangdong	88.6	5.80	88.9	6.22	97
Guangxi	86.5	5.39	83.6	3.75	88
Hainan	85.4	5.59	94.0	4.43	91
Sichuan	83.2	5.19	93.8	5.10	90
Guizhou	70.5	4.13	74.2	2.86	84
Yunnan	70.6	4.09	71.6	3.36	84
Xizang	38.5	1.82	25.4	2.06	N.A.
Shaanxi	81.7	5.53	98.6	9.00	94
Gansu	66.0	4.40	78.7	4.71	91
Qinghai	62.6	4.39	70.6	3.55	83
Ningxia	73.6	4.85	75.2	5.47	88
Xinjiang	86.6	5.70	86.0	8.65	92
China		**5.45**		**6.50**	**92**

Sources: *Tabulations of China's 1990 Population Census and 1995 one-per cent sample Population Survey*, published by China's State Statistical Bureau; *China Education Statistics Yearbook, 1995*, Published by China's State Education Commission, 1996.

Notes:
1. The literacy rate for 1995 is calculated for the population aged 15 and over.
2. Average years of schooling are calculated for the population aged 6 and over.
3. Percentage of enrolment for children aged 8 is calculated using 1990 Population Census data.
4. Data for 1994 refers to enrolment rate for people aged 18–21.
5. Data for 1990–92, from UNDP *China: Human Development Report, 1998*, p. 38.

education, which themselves result from the general trend towards decentralization of decision-making processes in both administration and the market (Tsui, 1997).

According to the 1990 census, there were 204.855 million illiterate people in China, accounting for 18.12 per cent of the total population. One third of them were aged 60 years and over. Considering the low age-specific illiteracy rate in younger age groups, China can be expected to move smoothly towards universal literacy as the younger cohorts gradually replace the older age groups. It should, however, be noted that illiteracy is not a problem occurring only among the oldest age group. There were 6.1 per cent of people aged between 15–29 in 1990 who were illiterate, indicating that the elimination of illiteracy is a very long and difficult task facing Chinese society.

In general, illiteracy rates tend to be lower in China's urban areas and in the eastern coastal regions, while the high illiteracy rates are concentrated in the west, and especially southwest part of China. Among the 181.61 million illiterate people aged 15 years and over, 15.16 per cent live in urban areas, while the remaining 84.84 per cent live in the countryside. The rates of illiteracy and semi-illiteracy have been declining in the long term in both urban and rural areas. Although the rural–urban gap is shrinking, there is concern over the negative impact of the economic reform on educational attainment in rural areas.

There are striking regional disparities (table 10.6). Adult literacy in Beijing, Shanghai, Tianjin, Liaoning, Jilin and Shanxi was above the 90 per cent mark, while the rate in Xizang was lower than 40 per cent. The regional disparities were largely driven by differences in female literacy. Female literacy rates are below male literacy rates, but the gender gap is narrowing.[5]

Based on provincial data for the educational attainment of the population aged 6 and over, average years of schooling were calculated for each provincial unit. For China as a whole, the average number of years of schooling was 5.45 years in 1990. There were seventeen provinces where the index was higher than the national average. Beijing and Shanghai were the places with the largest number of years of schooling (more than seven years). Among the thirteen provinces with the average years of schooling below the national average, the four provinces in the southwestern part of China, namely Xizang, Yunnan, Guizhou and Qinghai, recorded the lowest figures (see figure 10.2). These provinces also have both high illiteracy rates (see figure 10.3) and a low proportion of people with higher education.

The percentage of enrolment for children at age 8 can be used as an index to refer to the regional diversities in elementary education. As a result of different local traditions and practice, the age at which children start their formal education varies between regions (see figure 10.4). While nearly 100 per cent enrolment rates were achieved around age 8 in provinces such as Shanghai, Beijing and Tianjin, the percentages of children at age 8 enrolled

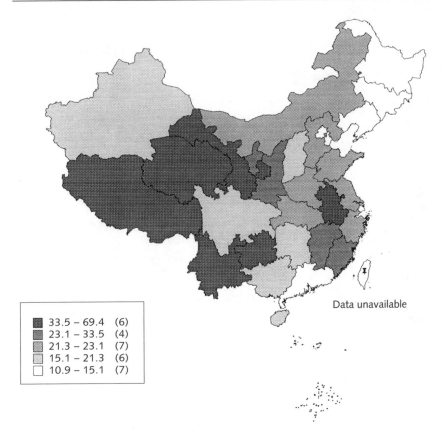

33.5 – 69.4   (6)
23.1 – 33.5   (4)
21.3 – 23.1   (7)
15.1 – 21.3   (6)
10.9 – 15.1   (7)

Data unavailable

Figure 10.3   Illiteracy rate, China, 1990

in primary schools was lower than 75 per cent in Xizang, Guizhou, Yunnan
and Qinghai in 1990.

Just as there is a marked regional difference in enrolment, there is similar
difference in graduation. Table 10.6 shows that while students in the eastern
provinces in general complete their primary education, only 79 per cent of
their counterparts in Jiangxi province do so. In Jiangxi, of every 1000
students in primary school, only 275 are able to enter junior school after
graduating.

As discussed earlier, investment in education, especially in elementary
education, comes almost entirely from the local resources. In 1990 direct
input by the provincial government was 290 yuan for each student in elementary education in Beijing, but only 61 yuan for each student in Hubei
Province. There are also regional variations within provinces. Many poor
areas have to collect higher fees from the parents to keep the local educa-

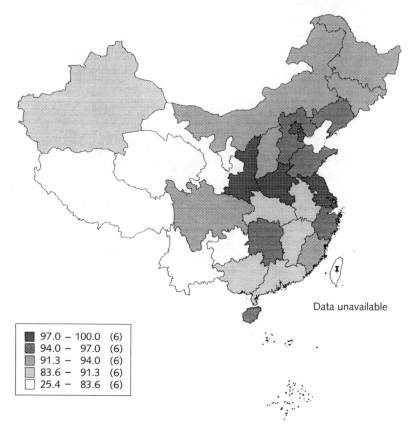

Data unavailable

■	97.0 – 100.0	(6)
▨	94.0 – 97.0	(6)
▩	91.3 – 94.0	(6)
▨	83.6 – 91.3	(6)
□	25.4 – 83.6	(6)

Figure 10.4   Percentage enrolment for people aged 8, 1990

tion system running. This in turn greatly restricts access to primary education in these areas, and results in a massive problem of illiteracy in the
younger age groups.

Notes

1. Details of the new strategy can be found in '*Mianxiang 21shiji jiaoyu zhenxing
   xingdong jihua* (Action Plan for Education Invigoration in the 21st century)',
   which was formulated by China's ministry of Education on 24 December 1998
   and approved by the State Council on 13 January 1999.
2. See *People's Daily* (English edition) 25 February 1999.
3. There is no uniform definition for school-age children. It varies across provinces.
   In some it refers to the age group 6–12, but in others to the 7–12 age group.
4. Information about the 'Hope Project' is basically from 'The Evaluation Report
   on the Hope Project' by the National Research Centre for the Promotion of

Science and Technological Development', 1998, Beijing. The report makes use of information collected from 695 rural counties. A condensed version of the report was published in *Zhongguo Qingnian Bao* (The China Youth), 30 October 1998. Interested readers can also visit its website: *www.projecthope.org*.
5. Details of gender difference in educational attainment are dealt with in Chapter 12 of this book.

References

Beijing Review (1994), 15 years of economic reform in China. Beijing: New Star Publisher.

Beijing Review (1999), Chinese education more connected with the outside world, 42 (5).

Brown, Hubert O. (1991), People's Republic of China. In Philip G. Altbach (ed.). *International Higher Education Encyclopaedia, Volume 1*, New York: Garland Publishing.

Dong, Chun-Cai (1988), The People's Republic of China. In T. Neville Postlethwaite (ed.), *The Encyclopaedia of Comparative Education and National Systems of Education*, Oxford: Pergamon Press.

Dreze, Jean and Loh, Jackie (1995), Literacy in India and China. *Economic and Political Weekly*, 30 (45), 2868–78.

Education Yearbook of China, The (1994–7), Beijing: China Education Publishing House.

Gao, Qi (ed.) (1985), *The History of Contemporary Chinese Education*. Beijing: Beijing Normal University Press.

Hayhoe, Ruth (ed.) (1992), *Education and Modernisation: The Chinese Experience*. Oxford: Pergamon Press.

Hu, Ruiwen (1996), *The New Development in China's Education during the 1980s and Prospect of the 1990s*. An internal publication of the Shanghai Institute for Intelligence Development Research.

Ministry of Education, China (1998), *Action Plan for Education Invigoration for the 21st Century*.

People's Daily (English edition) (1999), China to push forward education in qualities. 17 June 1999.

Pepper, Suzanne (1990), *China's Education Reform in the 1980s*. China Research Monograph, Institute of East Asian Studies, University of California at Berkeley.

State Education Commission of PRC (1996), *Statistical Communiqué on 1995 National Educational Development*. Beijing.

Tsui, Kai-yuen (1997), Economic reform and attainment in basic education in China, *The China Quarterly* 149, 105–28.

UNDP (1998), *China: Human Development Report, Human Development and Poverty alleviation, 1997*.

Yang, Tongping (1997), China entering the education century. In Zhao, Dingyang et al. *Ideas and Problems of China*, Nanchang, China: Jiangxi Education Publishing House.

11 China's Rural and Urban Employment in the Reform Era: Recent Trends and Policy Issues

Zuo Xuejin

One of the greatest challenges China has been facing in its socio-economic development is how to make a more efficient use of its abundant labour resources. A dominant concern has been the effect of population pressure on economic growth and employment, and this has been the primary motivation for the government's restrictive and somewhat controversial population policies, which have interfered with both fertility and mobility.

The lagged effect of high fertility in the 1950s and 1960s, encouraged by the government's pronatalist policy during that period, led to a rapid growth of the labour force in the 1970s and 1980s. This demographic change, coupled with a development strategy biased in favour of heavy industry, caused serious difficulties for the governments of urban localities in finding jobs for the growing number of urban entrants. During the Cultural Revolution (1966–76), the government initiated a campaign to rusticate urban dwellers, especially junior and senior high school students, to the countryside (*shangshan xiaxiang* in Chinese) to alleviate urban unemployment. Agriculture was used as a reservoir for the rural and urban surplus labour force. This practice, to a certain extent, only shifted the employment burden from urban to rural areas. Due to the declining per-worker holding of farmland and the lack of non-agricultural employment opportunities, this policy inevitably caused the employment situation in rural areas to deteriorate.

It was under this special set of circumstances that rural communities in some coastal provinces (e.g. Jiangsu Province) started to develop 'commune and brigade enterprises' (after the dismantling of the commune system in early 1980s, they were renamed 'township and village enterprises'). Although these rural enterprises were not encouraged, they were not actually discouraged by official policy in their early stage of development.

The campaign of rusticating urban youth was abandoned in the late 1970s. Fortunately, rural reform and strong rural and urban economic growth facilitated the absorption of the swelling labour force in the early 1980s. Several employment-promoting factors came into play, such as the dramatic growth of rural manufacturing, construction, transportation and other service sectors, increased inter-sectoral and inter-regional mobility of the rural labour, and the fast growth of urban economies, especially the non-state formal sectors and traditional services.

The growth of the working-age population has slowed down since the mid-1980s, primarily due to the lagged effect of fertility decline in the 1970s. Nevertheless, the deepening of urban enterprise reform and the deteriorating performance of state-owned enterprises (SOEs) and urban collective enterprises (UCEs) have led to an increased number of laid-off urban workers since 1993. The protectionist regulations in urban localities to restrict the employment of rural–urban migrant workers, a development strategy biased toward capital-intensive industries and induced by the artificially high urban wage rate, and the subsidized bank loans available to the urban formal sectors, have unfortunately deterred the efficient allocation of rural and urban labour resources.

This chapter intends to investigate the trends of labour supply and demand in rural and urban sectors and the relevant policy issues during the reform period. The first section will review the general trend of employment in China since the commencement of the reform in 1978. It will be followed by more detailed discussions of rural and urban employment, respectively. The fourth section will turn to an analysis of the major causes of urban unemployment and their policy implications. Concluding remarks are presented in the last section.

General Trends since 1978

The State Statistical Bureau (SSB) (1996) data report that China's total population increased from 963 million in 1978 to over 1,211 million in 1995, growing at an annual rate of 1.36 per cent. The growth rate declined significantly in the early 1990s, as the population growth averaged 1.16 per cent during 1990–95, compared with 1.37 per cent per annum during 1978–85 and 1.55 per cent during 1985–90. A further decline to below 1 per cent per annum during 1995–2000 is predicted.

The same source reports that China's total working-age population increased from 485 million in 1978 to 697 million in 1990, implying an average annual growth rate of 3.1 per cent, or more than double the growth rate of the total population. A large number of 'baby boomer' cohorts, born in the 1960s, inflated the working-age population by 3.6 per cent annually during 1978–85. As a result of sharp fertility decline in the 1970s, the growth rate of the working-age population declined to a moderate level of 2.3 per cent during 1985–90. Utilizing age-sex structure data from the 1982 and 1990 censuses of China, Banister (1992) projects that growth in the working-age population will average only 1.3 per cent per annum in the 1990s, and 1.1 per cent during 2000–10. Accordingly, the average additions to the working-age population will reduce by nearly half from 17.7 million annually during 1978–90 to 9.0 million during 1990-2000 (table 11.1).

Table 11.1 Trends of labour supply, 1978–2000

Year	Total population			Working-age population	
	Total size (million)	Percentage urban (%)[a]	Average annual growth (%)[b]	Total size (million)	Average annual growth (%)[b]
1978	962.59	17.9		485.3	
1985	1,058.51	23.7	1.37	621.1	3.6
1990	1,143.33	26.4	1.55	697.3	2.3
1995	1,211.21	29.0	1.16	na	
2000	1,270.00	33.0	0.95	790.0	1.3

Notes:
(a) Percentage urban is based on the classification between village committee (*cun wei hui*) and the resident committee (*ju wei hui*): all the residents in the former are reported as urban and otherwise reported as rural. However, temporary migrant population is likely to be under-reported in urban places in population census and surveys.
(b) Average annual growth between the year and the previous year reported in the table.
Sources: SSB (1996) p. 69 for total population and percentage urban up to 1995; author's estimate for total population in 2000; SSB (1993) for working-age population up to 1990; and Banister as cited by World Bank (1992) for working-age population in 2000.

In contrast to the deceleration in the growth of the working-age population since the mid-1980s, strong economic performance has favoured a fast growth in rural and urban employment. While the average growth rate of the working-age population (3.6 per cent) exceeded that of total employment (3.1 per cent) during 1978–85, the situation was reversed in the later years, as total employment grew at 2.6 per cent and 1.9 per cent per annum, respectively, during 1985–90 and 1990–95, surpassing the working-age population growth rate over the corresponding periods.

Consistent with the experiences of developed countries in their early stage of development and most other developing countries, China's economic development is accompanied by industrialization and urbanization, implying a faster growth in urban sectors than rural sectors. As reflected in the table 11.2, urban employment grew faster than rural employment over all the three reported periods. This is more apparent during 1990–5, when the growth rate of urban employment was more than double that of rural employment. As the pattern of demographic changes, sources of employment growth, and institutional arrangements are quite different in rural and urban areas, rural and urban employment is to be discussed separately in the following two sections.

Table 11.2   Trends of rural and urban employment, 1978–2000

	Total		*Urban*		*Rural*	
	Total employment (million)	*Average annual growth (%)*	*Total employment (million)*	*Average annual growth (%)*	*Total employment (million)*	*Average annual growth (%)*
1978	401.5	—	95.1	—	306.4	—
1985	498.7	3.1	128.1	4.3	370.7	2.8
1990	567.4	2.6	147.3	2.8	420.1	2.5
1995	623.9	1.9	173.5	3.3	450.4	1.4
2000	686.0	—	204.3	—	482.9	—

Sources: (SSB 1996) for 1995 and earlier years. Figures for 2000 are derived from author's estimates by assuming the same growth rate of rural and urban employment for 1995–2000 as that for 1990–5.

Rural Employment

When the reform started in the late 1970s, the rural population accounted for over 80 per cent of the total population, and the rural economy was primarily agricultural. About 250 million people, or one-third of the rural population, were living in absolute poverty (World Bank, 1992). As a result of fast population growth and the lack of non-agricultural employment opportunities, underemployment was a prevailing problem in rural areas. Rural–urban migration was generally prohibited under the institutional arrangements for rural-urban segmentation enforced by the household registration system and the grain-food rationing system.

Reform has accelerated the transfer of agricultural labour to non-agricultural activities by creating increased employment opportunities in local rural enterprises and by encouraging migration away from home areas. Total rural employment increased from slightly over 300 million in 1978 to 450 million in 1995, at an annual rate of 2.3 per cent over the 17-year period. However, there was a general trend of slowdown in the growth of total rural employment, as the average growth rate fell from 2.8 per cent during 1978–85 to 2.5 per cent during 1985–90, and, more significantly, to 1.4 per cent during 1990–5. Two factors are primarily responsible for this slowdown in total rural employment. It might, in the first place, simply be a reflection of a slowdown in the rural labour supply; official data do not allow overt rural unemployment, and hence tend to equate rural employment with rural labour supply by including substantial rural underemployment in the employment figure. Both fertility decline in the 1970s and accelerated urbanization since the reform have, in turn, contributed to the slower growth of rural labour force. In the second place, the slowdown doubtless has something to do with the sluggish and eventually negative growth of agricultural employment. Agricultural employment has experienced a continuous decline since reaching its peak in 1991. During 1992–5, agricultural employment declined by 17 million, meaning that non-agricultural sectors had to create 5–6 million new jobs per year merely to offset the decline in agricultural employment and to prevent total rural employment from declining.

Fortunately, the economic boom during the period 1992–5 brought strong growth in rural non-agricultural sectors. Non-agricultural employment increased by nearly 30 million, resulting in a nearly 13 million net increase in total rural employment. As a result of differential growth in the agricultural and non-agricultural sectors, the share of agricultural employment dropped from nearly 90 per cent of total rural employment in 1978 to about 72 per cent in 1995.

As mentioned earlier, official data treat agricultural employment as the residue of the rural labour force net of non-farm employment, and hence no overt unemployment can be found in rural sectors. Furthermore, the official employment data are based on people's major occupation, and hence tend to ignore the seasonal and temporary activities of rural households. After analysing the available data on average labour input per acreage of farmland and per head of cattle, Rawski and Mead (1997) argue that China's official data systematically overstate the country's agricultural employment by underreporting off-farming local employment and employment through out-migration.[1]

Non-agricultural employment

Official data show that the non-agricultural sector accounted for over 28 per cent of total rural employment in 1995, and it has replaced agriculture to become the major and even the only source of rural employment growth since early 1990s. During 1985–90, non-agricultural sectors accounted for 40 per cent of the total absorption of 50 million rural labourers; the figure increased to 133 per cent of the 30 million absorption between 1992 and 1995, as part of the increase was offset by the decline in agricultural employment. In 1995 for every two jobs created by the non-agricultural sector, one was used to offset the decline in farm employment, and the other for the net increase in the rural jobs.

The rural non-agricultural sector consists of several subsectors, including industry, construction, transportation and communication, wholesale and retail trade and restaurants, and other non-agricultural services. The role of rural industry in rural labour absorption declined in the decade 1985–95, as it provided less than 30 per cent of the non-agricultural employment in 1995, compared to over 40 per cent in 1985. Rural services grew significantly faster than rural industry. However, it is likely that, under the current statistical reporting system, some rural services are actually performed in urban areas, since the employment of construction workers and

Table 11.3 Rural employment by sector (in millions), 1978–1995

	Total	Agriculture	Non-agriculture	Industry	Construction	Transportation	Trade, restaurants and services	Other NA
1978	306.4	n.a.	n.a.	n.a.	n.a.	n.a.	n.a.	n.a.
1980	318.4	n.a.	n.a.	n.a.	n.a.	n.a.	n.a.	n.a.
1985	370.7	303.5	67.1	27.4	11.3	4.3	4.6	19.5
1986	379.9	304.7	75.2	31.4	13.1	5.1	5.3	20.4
1987	390.0	308.7	81.3	33.0	14.3	5.6	6.1	22.3
1988	400.7	314.6	86.1	34.1	15.3	6.1	6.6	24.1
1989	409.4	324.4	85.0	32.6	15.0	6.1	6.5	24.7
1990	420.1	333.4	86.7	32.3	15.2	6.4	6.9	25.9
1991	430.9	341.9	89.1	32.7	15.3	6.6	7.2	27.3
1992	438.0	340.4	97.6	34.7	16.6	7.1	8.1	31.2
1993	442.6	332.6	110.0	36.6	18.9	8.0	9.5	37.0
1994	446.5	326.9	119.6	38.5	20.6	9.1	10.8	40.6
1995	450.4	323.3	127.1	39.7	22.0	9.8	11.7	43.8
Average annual growth rate								
1978–85	2.8							
1985–90	2.5	1.9	5.3	3.3	6.1	7.9	8.4	5.9
1990–95	1.4	−0.6	7.9	4.2	7.7	9.1	11.0	11.1

Source: SSB (1996), *Statistical Yearbook of China, 1996.*

the informal employment of migrants in urban areas are usually recorded at their rural places of origin rather than at their urban destination. This is also reflected in the growing figure for 'other non-agricultural employment' in the last column of the table: at least one major component of this category may be the employment of rural out-migrants. Given the limited information available about the out-migrants' activities at their destinations, it is easier for officials at their places of origin to report the out-migrants' activities as 'other non-agricultural employment'.[2]

Development of rural enterprises

The rapid expansion of rural enterprises, which are predominantly non-agricultural, characterizes the post-1984 development of China's rural economy. Table 11.4 contains information about employment in rural enterprises (REs). Total employment in REs increased from 28.3 million in 1979 to 128.6 million in 1995, growing at an average annual rate of nearly 10 per cent. Sub-village REs, including those collectively owned by the sub-village units (formally production teams) and private enterprises, appear to be more vital than township and village enterprises, as employment in the former expanded by more than 41 million during 1985-95, providing over 70 per cent of the RE incremental employment. Their share of RE employment surpassed that of TVEs for the first time in 1990. As regards the sectoral distribution of RE employment, industrial employment has stabilized at about 60 per cent of total RE employment since 1985, followed by employment in services (about 37 per cent), whereas the share of agricultural employment has been negligible.

Evidently, RE employment is one of the most important opportunities that the reform has offered to Chinese peasants. Without this opportunity, the overall picture of rural employment, the rural economy and the national economy would all be substantially altered.

However, there have been substantial regional disparities in the development of REs. RE performance has been highly correlated with county and township income levels. A comparison of the three macroeconomic regions finds that development of REs in the remote and poor western provinces has been quite limited as compared to development in the prosperous coastal provinces. In 1995, for instance, RE employment accounted for 39 per cent of total rural employment in the coastal belt, below 28 per cent in the central belt, and just over 17 per cent in the western belt. The provinces with the lowest percentage of RE employment were the poor northwestern and southwestern provinces of Guizhou, Guangxi, Yunnan, Qinghai and Ningxia. In the 1980s, the REs in the western belt tended to grow faster, as the belt's growth rate of nearly 15 per cent was higher than the 13 per cent in the central belt and 10 per cent in the coastal belt. However, this advantage in growth rate of the western belt faded in the first half of 1990s, as all the three macro belts grew at quite similar rates in the

Table 11.4 Employment of rural enterprises, 1979 and 1984–1995

Years	Total employment (million)	By ownership (%)		By sector (%)		
		Township and village	Subvillage	Agriculture	Industry	Services
1979	28.3					
1984	52.1	76.4	23.6	5.5	70.2	24.4
1985	69.8	62.0	38.0	3.6	59.3	37.1
1986	79.4	57.2	42.8	3.0	60.0	37.0
1987	88.1	53.6	46.4	2.8	59.8	37.4
1988	95.5	51.2	48.7	2.6	59.7	37.7
1989	93.7	50.4	49.6	2.6	60.0	37.4
1990	92.6	49.6	50.4	2.5	60.2	37.3
1991	96.1	49.6	50.4	2.5	60.5	37.0
1992	105.8	48.7	51.3	2.4	59.9	37.7
1993	123.5	46.7	53.3	2.3	58.8	38.9
1994	120.2	49.0	51.0	2.2	57.9	39.9
1995	128.6	47.2	52.8	2.4	58.8	38.7

Source: SSB, 1996.

neighbourhood of 7 per cent. Furthermore, during this period, the poorest provinces, Guizhou, Guangxi, Qinghai and Ningxia experienced negative growth in RE employment, leaving them farther behind the others.

Clearly, villages in the remote, mountainous provinces are disadvantaged in developing REs due to their poor endowment in human capital and financial resources, and difficult access to roads and domestic and international markets. Owing to the fairly limited opportunities for local off-farming employment in poor areas, out-migration to prosperous areas has, since the early 1980s, become another option for the poor seeking better job opportunities.

Rural out-migration

In spite of tight government control over permanent rural-urban migration, the scale and significance of temporary rural-urban migration has been increasing since the early 1980s. An array of demographic and economic factors have contributed to the increase in rural labour mobility from the poor to prosperous and from rural to urban areas. On the supply side, earlier and more profound urban fertility decline in the past decades has led to slower growth of the urban-born working-age population than of rural working-age population; on the demand side, urban employment has grown faster than rural employment since 1978, especially since the

Table 11.5 Provincial rural enterprise employment, 1980–1995

	RE employment (000s)			% rural working population			Growth rate (%)	
	1980	1990	1995	1980	1990	1995	1980–90	1990–95
Total	29,997	92,647	128,621	9.4	22.1	28.6	11.9	6.8
Coastal Belt								
Subtotal	17,078	45,927	63,707	14.1	29.6	39.0	10.4	6.8
Beijing	314	1,088	987	19.1	59.1	60.4	13.2	−1.9
Tianjin	302	887	1,047	20.0	51.9	61.6	11.4	3.4
Hebei	1,886	6,412	8,520	10.7	27.2	33.1	13.0	5.8
Shandong	3,343	9,439	14,398	12.6	28.2	40.3	10.9	8.8
Liaoning	1,117	3,051	4,342	16.5	36.7	50.1	10.6	7.3
Shanghai	699	1,515	1,399	24.8	60.6	60.7	8.0	−1.6
Jiangsu	3,886	8,961	9,247	17.2	32.2	33.3	8.7	0.6
Zhejiang	2,349	4,955	7,957	15.7	24.3	37.9	7.7	9.9
Fujian	1,174	2,791	4,710	16.1	27.6	41.0	9.0	11.0
Guangdong	2,008	6,828	11,100	10.2	26.7	40.8	13.0	10.2
Central Belt								
Subtotal	8,837	30,396	41,919	8.6	21.6	27.5	13.1	6.6
Heilongjiang	485	1,622	2,095	19.1	30.1	35.9	12.8	5.3
Jilin	378	1,552	2,138	11.2	25.3	33.9	15.2	6.6
Henan	1,603	8,819	7,160	6.4	25.8	19.0	18.6	−4.1
Inner Mongolia	265	976	2,240	6.0	18.1	38.0	13.9	18.1
Shanxi	883	2,404	4,073	12.6	27.1	43.0	10.5	11.1
Anhui	864	4,623	5,950	5.2	20.1	23.0	18.3	5.2
Jiangxi	838	2,327	4,401	9.0	16.7	28.3	10.8	13.6
Hubei	1,527	3,890	6,636	10.2	21.7	36.7	9.8	11.3
Hunan	1,994	4,183	7,226	10.0	16.1	26.2	7.7	11.6
Western Belt								
Subtotal	4,082	16,278	22,995	4.5	13.2	17.2	14.8	7.2
Shaanxi	628	2,536	3,290	7.0	21.1	24.8	15.0	5.3
Ningxia	51	235	143	5.1	16.7	8.8	16.5	−9.5
Gansu	226	1,294	2,043	3.9	15.8	23.3	19.1	9.6
Qinghai	38	146	93	3.7	10.8	6.1	14.4	−8.6
Xinjiang	194	464	500	7.5	15.9	16.0	9.1	1.5
Guangxi	593	1,991	1,828	4.6	11.3	9.3	12.9	−1.7
Sichuan	1,730	7,058	11,507	4.6	14.4	22.2	15.1	10.3
Guizhou	201	1,069	857	2.2	7.6	5.4	18.2	−4.3
Yunnan	421	1,485	2,734	3.5	9.0	14.9	13.4	13.0
Tibet		44			4.9			

Notes:
(a) Hainan was part of Guangdong before it becoming an independent province in early 1980s. To make the inter-temporary comparison, Hainan is included in Guangdong for 1990 and 1995 figures.
(b) Tibet is not included in the calculation due to inadequate data.
Source: SSB, *Statistical Yearbook of China, 1995* and earlier versions.

mid-1980s. This trend is likely to continue in the 1990s. Inter-regional and rural-to-urban income disparities have increased since the mid-1980s, providing peasants with strong economic incentives to migrate. In addition, the government has relaxed some restrictions on rural labour mobility since the early 1980s, which has also promoted the migration.

Nevertheless, information about the scale of rural–urban migration has remained limited, since the floating population of temporary migrants who have been away from their place of registration for less than one year, has never been counted in the census so far.[3] The 1990 census recorded a total number of 34.1 million long-term migrants, accounting for 3 per cent of the 1.13 billion total population. Assuming rather conservatively that an equal number of temporary migrants were not recorded in the census, then the total number of migrants should be doubled to 68 million in mid-1990. Many other researches suggest that the total volume of these temporary migrations exceeds 80 million.

Cities in the prosperous coastal provinces have become the import-receiving places of rural migrants. For instance, the 1993 survey of the floating population in Shanghai revealed that it totalled 2.5 million, or about one-fifth of the local registered population. In some of the newly developed cities of Guangdong, such as Shenzhen, Nanhai and Shunde, migrants account for over 50 per cent of the local 'registered' population.

Rural–urban migration is an important channel for peasants in poor inland provinces to participate in the fast economic growth in coastal areas. A 1993 survey in 50 townships across the country found that peasants in the western provinces had a higher propensity to migrate (Zhang Qingwu, 1995). Several studies found that out-migration had played an important role in the income growth of rural households in poor areas (Wang and Zuo, 1996).

Nevertheless, in-migration of peasant workers has concerned local governments and residents in the cities. They are concerned with the consequences of increased rural-urban migration, such as the overcrowding of urban infrastructures, an increased urban crime rate, and most of all, higher urban unemployment. In many urban areas these concerns have already been translated into migration-restricting measures such as regulations to limit the employment of migrant workers and to impose new taxes or fees on the migrants and their employers.

Urban Employment

The trend in urban employment

Urban employment increased from 95 million to 174 million over the period 1978–95, growing at an average annual rate of 3.6 per cent, much higher than the rural rate of 1.4 per cent during the same period.[4] Within the urban economy, the state, collective and private (plus self-employed) sectors provided 78.3 per cent, 21.5 per cent and 0.3 per cent of total employment in 1978. As a result of the much higher growth rate of private sectors than the

Table 11.6   Urban employment by ownership of the employer (in 10,000s)

Year	Total	State		Collective		Private		Self-employed	
		Number (10,000)	(%)	Number (10,000)	(%)	Number (10,000)	(%)	Number (10,000)	(%)
1978	9,514	7,451	78.3	2,048	21.5	0	0.0	24	0.3
1980	10,525	8,019	76.2	2,425	23.0	0	0.0	15	0.1
1984	12,229	8,637	70.6	3,216	26.3	37	0.3	81	0.7
1985	12,808	8,990	70.2	3,324	26.0	44	0.3	450	3.5
1986	13,293	9,333	70.2	3,421	25.7	56	0.4	483	3.6
1987	13,783	9,654	70.0	3,488	25.3	72	0.5	569	4.1
1988	14,267	9,984	70.0	3,527	24.7	97	0.7	659	4.6
1989	14,390	10,108	70.2	3,502	24.3	132	0.9	648	4.5
1990	14,730	10,346	70.2	3,549	24.1	164	1.1	671	4.6
1991	15,268	10,664	69.8	3,628	23.8	216	1.4	760	5.0
1992	15,630	10,889	69.7	3,621	23.2	282	1.8	838	5.4
1993	15,964	10,920	68.4	3,393	21.3	536	3.4	1,116	7.0
1994	16,816	11,214	66.7	3,285	19.5	759	4.5	1,557	9.3
1995	17,346	11,261	64.9	3,147	18.1	894	5.2	2,045	11.8
Growth									
1980–95	3.4	2.3		1.8		33.6a		38.8	
1980–5	4.0	2.3		6.5		18.9b		97.4	
1985–90	2.8	2.8		1.3		30.1		8.3	
1990–5	3.3	1.7		–2.4		40.4		25.0	

Notes:
(a) Growth rate for 1984–95.
(b) Growth rate for 1984–5.
(c) Private sectors include foreign-funded, joint-venture, and private-owned enterprises.

state and collective sectors, these shares changed significantly between 1978 and 1995. In 1995 they provided, respectively, 65 per cent, 18 per cent and 17 per cent of total urban employment. During 1990–5 employment in the state sector grew at 1.7 per cent per annum; however, most of the growth came from employment in government agencies and public institutions, whereas employment in the state enterprises actually declined.

Official statistics for urban employment are dominated by the formal employment of urban 'workers and staff' (*zhigong*), and they neglect the often temporary or informal employment of migrant workers. Therefore, official data generally tend to under-estimate the actual size of urban employment. For instance, there are more than 2 million migrants working in Shanghai, accounting for over 40 per cent of the employment of 'workers and staff'.

Urban unemployment and lay-offs

Urban unemployment has increasingly been an issue since 1992, when the state-owned enterprise (SOE) reform proceeded to tackle the difficult problem of over-staffing and deteriorating performance in the SOEs. In fact, the official urban unemployment rate increased only slightly over 1985–95, from less than 2 per cent to 2.9 per cent. However, these unemployment figures do not include those laid-off workers, who have no work to do but receive some living expenses from their employer units. If these laid-off workers were included, then urban unemployment would be much higher. The government authorities revealed that by the end of the first quarter of 1997, nearly 11 million formal employees (staff and workers) in the SOEs and urban collective enterprises of the country received reduced or no remuneration, and about 2.3 million retirees received reduced or no pensions. Over 9 million employees were laid off, reaching the peak towards the end of the 1990s. In 1996 (Wu Bangguo, 1997) the number of laid-off workers in Shanghai alone totalled about 200,000. By comparison, the number of unemployed amounted to about 150,000. In some provinces with a large concentration of SOEs, such as the northeastern provinces of China, the laid-off rates are much higher.

Owing to the political sensitivity of urban unemployment, the issue has attracted considerable attention from both researchers and governments at the central and local level. Government responses to increased urban unemployment include tightening the restrictions on the employment of migrant workers in cities; income maintenance and re-employment programmes targeting laid-off workers, etc.

In Shanghai, for instance, the government has set up a number of 're-employment centres' to accommodate laid-off workers whose employer units can not pay their living expenses and benefits. These centres are financed by government, businesses and non-governmental organizations. Their basic function is to continue payments to laid-off workers, to retrain them and to help them find new jobs.

However, to make the government programmes more effective, it is of critical importance to identify the major causes of urban unemployment correctly. This will be the focus of discussions in the next section of this chapter.

Major Causes of Urban Unemployment

Some scholars argue that government anti-inflation measures, such as the contractive monetary policy, are responsible for the high urban unemployment rate. They suggest that a lower urban unemployment rate can be achieved at a higher inflation rate.

The theoretic foundation of this argument is the so-called 'Philip's Curve', that describes a negative correlationship between the two major macroeconomic aims of the government, that is, fighting inflation and fighting unemployment. However, this hypothesis has not received much support from empirical evidence. Furthermore, high urban unemployment has occurred since 1992, when the economy was already growing at a double-digit rate. While most economists are concerned with the sustainability of such growth, the statement that attributes urban unemployment to contractive monetary policy is obviously questionable.

Many local workers and local governments in the receiving areas tend to believe that the high unemployment (lay-off) rate is caused by the increased flow of migrant workers into the cities. In other words, laid-off local workers are replaced by migrants. Based on this hypothesis, local governments have instituted more restrictive regulations to limit the scope for the employment of migrants in urban sectors, and to discourage the hiring of migrants by local enterprises.

This common-sense belief, however, can be misleading. A cross-sectional analysis of the urban unemployment rate does not find a positive association between the local unemployment rate and the in-migration rate. In fact, urban areas with the highest in-migration rates, such as Shanghai, Guangzhou and Shenzhen, tend to have the lowest urban unemployment rates. On the other hand, urban areas with high unemployment rates are likely to have a lower in-migration rates. At the micro level, most lay-offs occur at money-losing enterprises, which are less likely to hire migrant workers.

It is commonly believed that the increase in the urban lay-off and unemployment rates has occurred as a result of the structural adjustment of urban economies. This is generally true. In coastal provinces, such as Shanghai, labour-intensive industry (e.g. the textile industry) has been declining in the past decades. Many textile mills have been closed, causing the laying-off of hundreds of thousands of textile workers. Some other industries, such as appliance and electronic instrument manufacturing, are facing the similar problem. However, the 'structural change' involved when an urban worker shifts from a textile mill to another urban job should be much easier to cope with than when peasants move from farming to urban jobs. Why is it that over 2 million migrants have found jobs in Shanghai, while 200,000 laid-off workers can not – despite systematic government efforts? Obviously, structural adjustment is not a satisfactory answer to the question.

The current high urban lay-off and unemployment rate is, to a large extent, caused by the totally different institutional arrangements for rural and urban residents. The high wage rates and generous benefits laid down by institutions lead to high labour costs for local workers. Based on a survey of the migrant population and a survey of local residents in Shanghai conducted in 1995 by the Institute of Population and Development Studies of the Shanghai Academy of Social Sciences, a regression analysis of personal income determination for both local and migrant workers finds that, controlling for human capital and occupation etc., local workers tend to have incomes 50 per cent higher than migrant workers. If we further distinguish local workers into 'local professional', 'local staff', 'local self-employed' and 'local manual worker', then we find that the largest income gap (about one-third) exists between local manual workers and their migrant counterparts, while the local self-employed have no advantage over their counterparts at all.

As this income difference can not be explained by a difference in human capital and productivity, it can be regarded as a 'rent' for 'urban job rights' or 'urban privileges'. Therefore, the urban wage (including housing, pension and health care benefits) can be broken down into two components: first, the competitive wage rate, which is determined by demand-supply forces in the labour market, and can be measured against the wage rate of a migrant worker of the same quality; second, the income which represents the rent for 'job rights' or 'urban privileges'. The cost of an institution-determined high urban wage is the high urban lay-off and unemployment rate. In fact, as revealed by the regression analysis, local manual workers are the most protected group, and they coincidentally have the highest unemployment rate. Moreover, very few local workers are engaged in self-employed activities, as urban job rights can not be realized in these activities.

The high lay-off and unemployment rates are the direct result of an institution-determined high urban wage rate. This is a legacy of the past. As China's reform has been based on Parato improvement, it would be difficult for the government to eliminate the urban privilege altogether. One possible solution to the problem is to pay urban workers competitive wage rates with appropriate compensation for urban 'job rights', which will allow urban workers to remain better-off as before, meanwhile removing the institutional barriers to the employment of migrant workers. This will lower the expected wage rate of urban workers, and hence increase their employment rate. Once the compensation for urban 'job rights' is ensured, they will be more willing to be engaged in small business, which will be an important source of urban employment growth.

Concluding Remarks

China's abundance in labour resources will not continue forever, instead, it will most likely come to an end somewhere between 2010 and 2030, when the 'baby boomers' born in the 1950s and 1960s age and retire. By coincidence, they are also the cohorts with unprecedentedly small families (one-child families for most urban couples) as they first experienced the fertility decline in their childbearing years during the 1970s and 80s. Therefore, it is of great importance that in the next two decades or so, China should make a better utilization of her labour resources, and prepare for the transition from labour-intensive technology to technology/capital-intensive technology. More attention should be paid to the employment impacts, both direct and indirect, of macroeconomic policies.

As the growth of the work force in rural areas will continue to be much higher than in urban areas, and as agricultural employment will continue to decline, the improvement of rural employment will heavily rely on both the development of local TVEs, the urbanization of industrialized rural areas, and rural out-migration. Since labour in the coastal provinces is becoming more expensive, both capital mobility from coastal to hinterland provinces and labour mobility in the opposite direction should be encouraged by the relevant policies. Any policy changes in this direction will not only help enhance labour allocation efficiency, but also contribute to an equal distribution of income and to the efforts of poverty alleviation.

As a much slower growth of the urban-born labour force coincided with strong growth in the economy, it is not correct to say that the major cause of high urban unemployment rate is an 'excess supply of urban workers'. Rather, it is the institutional segmentation of urban labour markets, and institution-determined high urban wage rates that cause high urban unemployment. Given the presence of market competition between the urban and rural enterprises in the product markets, any policy attempts to restrict rural-urban migration will not help resolve the problem. One feasible strategy for further reform is to compensate the job rights of urban workers, meanwhile removing the institutional barriers to the employment of migrant workers.

Notes

1. Rawski and Mead (1977) find that the official agricultural data exceed their reconstructed data by over 100 million, if the figure of 269 work days per year is applied to convert man-equivalents to person-years. If 300 days per year is used for the conversion, the gap is even larger.
2. The survey found that long-term (six months or longer) migrants accounted for 70 per cent of total migrants (Wang, Le and Zuo, 1995).
3. The 1990 Census for the first time collected information about change of residence in the past five years. However, only those who had stayed at their current residence or left their place of household registration for one year or longer were interviewed at the destination. Therefore, temporary and seasonal migration was not captured in the Census. In the 1987 and 1995 one-per-cent population surveys, the criterion of six months rather than one year was used.
4. The urban employment figures here may be under-reported, since they do not include the employment of migrant workers, which is officially reported at their rural places of origin.

References

Banister (1992), Background paper prepared for the World Bank.
Institute of Rural Development Studies at CASS, Department of Socio-Economic Statistics at State Statistical Bureau (1996), *Annual Report on China's Rural Economic Development and Analysis on the Development Trend in 1996*. Beijing: Publishing House of CASS.

State Statistical Bureau (1996), *Statistical Yearbook of China*, 1996 and the earlier years. Beijing: China Statistical Publishing House.

Wang, Wuding, Le, Huizhong and Zuo, Xuejin (eds) (1996), *Shanghai's Floating Population in 1990s*. Shanghai: The East China Normal University Press.

World Bank (1992), China: Poverty Reduction Strategy in 1990s. The World Bank.

12 China's Female Population

Tan Lin and Peng Xizhe

Introduction

Since the 1950s the role of Chinese women in society has been considerably transformed and the saying 'women hold up half the sky' has been taken seriously. The pursuit of gender equality has been regarded by the State as one of its central political goals. The Constitution of the People's Republic of China stipulates that 'women enjoy equal rights with men in all spheres of life, political, economic, cultural and social, including family life'. In addition to that, the Marriage Law, the Inheritance Act, the Electoral Law, the Criminal Law, and the Compulsory Education Law have also made detailed stipulations with regard to women's rights in the relevant areas and worked out measures for the protection of those rights. In April 1992 China's National People's Congress adopted the Law on the Protection of Women's Rights and Interests that consisted of 9 sections and 54 articles. This law provides a comprehensive guide to and protection of Chinese women's legal rights with regard to political and social life, culture and education, work, property, marriage and family and so on.

There are not only laws and regulations that protect women's basic rights, but also many social and economic measures that are particularly made for the purpose of gender equality. Consequently, women's status in China, especially in urban China, is high by international standards (Greenhalgh, 1991). Chinese women are better off, in terms of their health, status in the labour force and political participation, than women in other developing countries. According to the UNDP estimate, China is ranked number 40 among all countries in the world, based on the gender-empowerment measure (GSM), but only ranked 79 if another index, the gender-related development index (GDI), is used. The GDI value for China in 1997 was 0.699, higher than the average for the developing countries (0.630), but slightly lower than that of East Asia (0.709), and much lower than that of the industrialized countries (0.915) (UNDP, 1999). There are, nevertheless, large regional and rural-urban differences. Even within the urban areas, there exists a marked difference between women who hold permanent urban registry and those who do not.

It is well known that China is the most populous country in the world. According to the 1995 One-per-cent Population Sample Survey of China, (with 00:00 hour 1 October 1995 as the standard time), the mainland of China, comprising 30 administrative areas at provincial level, had a female population of 607,205,400 – 49.10 per cent of the total. The survey data revealed that the eastern coastal part of China had 41.47 per cent of the total female population, while the central and western regions accounted for 35.59 and 22.94 per cent, respectively. From the same source, we find that women aged 15–49 made up 55.52 per cent of the total female population, and older women aged 60+ accounted for 10.71 per cent. In 1995, of the total female population, 20.84 per cent held urban household registration records, 8.03 per cent of them were living in towns, and 71.13 per cent were in rural areas.

Those interested in China's population and demography inevitably need to study China's female population. This chapter focuses on profiling the female population during the last two decades, a period of profound socioeconomic reformation and transformation in China.

Women's Political Participation Situation since 1980

Economic reform and the open-door policy have brought tremendous changes in the political system of China, which have had a great effect on women's participation in politics. In the 1980s, with the progress of economic reform and government restructuring, the percentage of women in political and governmental leadership declined. This provoked a heated debate on the issue of women's political participation. Some argued that this decrease in the percentage of women leaders at each level was not an entirely negative phenomenon, as it marked a transition from women achieving political positions through their personal connections to women competing independently for posts. Others viewed this trend as a sign of the deterioration of the general position of Chinese women in society. One of the measures taken by the Chinese government to enhance women's political power was to compel every government agency to recruit a certain proportion of female staff. This measure remains in effect at present. Consequently, the rate of women's political participation has started to rise again. Table 12.1 shows us the percentage of female leaders among the total leadership working in government departments or Party or other organizations.

In 1993 female deputies accounted for 20 per cent of the total of the National People's Congress and 23.3 per cent of the members of its Standing Committee. In the same year, women made up 32.44 per cent of the total staff of government offices. Since the Fourth UN International Conference on Women and Development in 1995, the Chinese government has

Table 12.1 Percentage of female government or Party leaders in China

Year	Women leaders in government, Party or other organizations
1982	10.38
1987	12.16
1990	11.51
1995	14.55

Sources:
1982: *Population Census of China*, China Statistics Press, 1983.
1987: *1% Population Sample Survey*, China Statistics Press, 1988.
1990: *Population Census of China*, China Statistics Press, 1993.
1995: *1% Population Sample Survey*, China Statistics Press, 1997.

Table 12.2 Comparison of gender empowerment measure (GEM) for selected countries, 1997

Countries	Rank in human development index	Gender empowerment measure value rank	Seats in parliament (as % of total)	Administrators and managers (as % of total)	Professional and technical workers (as % of total)	Women's real GDP per capita (PPP$)
China	98	0.49140	21.8	11.6	45.1	2,485
Norway	2	0.81010	36.4	30.6	58.5	20,872
USA	3	0.70880	12.5	44.3	53.1	23,540
Japan	4	0.49438	8.9	9.3	44.1	14,625
Korea	30	0.33678	3.7	4.2	45.0	8,388
Hungary	47	0.45848	8.3	32.8	60.9	5,372
Brazil	79	0.36770	5.9	17.3	63.3	3,813
S. Africa	101	0.58218	28.4	17.4	46.7	4,637
India	132	0.24095	8.3	2.3	20.5	902

Source: UNDP, *Human Development Report*, 1999.

been committed to facilitating women's political participation, especially in the field of decision-making and management of state and social affairs. However, compared with many other countries, the proportion of Chinese women who participate in parliament or hold top government positions remains relatively low. Women's participation is high at the grass-root level, but rapidly declines as one rises up the political or administrative hierarchy. This is also generally true of other aspects of social life in China.

There are two major problems relating to women's participation in politics. The first is a lack of awareness of women's political participation among both women and men in Chinese society. The second is sex discrimination in the society as a whole. China was a male-dominated society for thousands of years, and women were never allowed to play an equal part in decision-making at any level. In the last several decades, especially in 1980s and 1990s, the Chinese government has made considerable efforts to safeguard equal legal rights for women in various aspects of society (see table 12.3). Nevertheless, changing traditional thinking about women is a huge challenge, and there is a need to call on the whole of society to respect and support women's empowerment, and particularly to encourage women's political participation.

Female Education

Female educational is a basic indicator of women's status in society. Since the founding of the People's Republic of China, it has been stipulated by law that women and men are equal and thus women should enjoy equal rights with men to be educated. Since the 1980s, tremendous changes have taken place in female education. They can be summarized as follows: the illiteracy rate has been significantly reduced; the average level of education has risen; and there has been a gradual narrowing of the differences in attainment between women and men as well as a gradual narrowing of the differences between urban and rural areas. However, female education levels vary and are affected by particular regional conditions, especially with respect to traditional and cultural background and uneven economic development.

Since the early 1980s the percentages of female students at each level has been increasing (see table 12.4). This was particularly notable in higher education where the percentage increased from 23.4 per cent in 1980 to 33.7 per cent in 1992, a total of 10.3 percentage points. However, the educational status of the female population as a whole is still very low, especially as regards higher education. Up to 1990 less than 5 per cent of women received higher education, and most females had not completed standard secondary education.

As shown in table 12.5, a severe degree of illiteracy still exists among the female population, for historical reasons and because of other social problems. As mentioned above, the female population is about the same size as the male population, but the number of illiterate females is much higher than that of males. In 1982 about half of all females in China were illiterate. The proportion decreased to 18.59 per cent in 1990, but fluctuated thereafter. One can also see from table 12.6 the female illiterates always form the major portion of total illiterates. The proportion increased from 69.21 per cent in 1982 to 72.63 per cent in 1995.

According to a survey on Women's Social Status in China conducted by the All-China Women's Federation and the State Statistics Bureau in 1990,

Table 12.3 Some important measures concerning women's rights in China since 1980

Year	Measure
1980	1. Marriage Law modified, and a new one promulgated. 2. Delegation organized to attend the second World Women's Conference held in Copenhagen. At this conference, the Chinese government signed the Convention Eliminating Discrimination Against Women in any Form.
1982	Constitution modified, emphasizing gender equality, especially women's social rights concerning participation in the process of policy and decision making in political, economic and family life.
1985	1. Governmental delegation organized to attend the third World Women's Conference. 2. Inheritance Law promulgated, clearly declaring that women shall enjoy equal inheritance rights with men. 3. Quality Standards and Requirements for Maternal Care in both urban and rural China released.
1986	1. Civil Law promulgated declaring that women shall enjoy equal civil rights with men. 2. Compulsory Education Law declaring that females shall enjoy equal education rights and opportunities with men. 3. Release of the Regulation covering Women and Child-Care Work.
1988	1. Protection Regulations for Female Workers and Employees released. 2. Regulation covering Maternity Treatment of Female Employees and Workers released.
1992	Law of The People's Republic of China on the Protection on Rights and Interests of Women promulgated.
1993	Goals of Chinese Women's Development in 1990s released.
1994	Labour Law promulgated, clearly declaring that women shall enjoy equal employment rights with men and women shall be protected during special periods, such as menstruation, pregnancy and breastfeeding.
1995	1. Fourth World Women's Conference held in China. 2. Programme for the Development of Chinese Women (1995–2000) released.

Source: Tan Lin and Li Xinjian (1995), *Women and Sustainable Development*, Tianjin Science and Technology Press.

Table 12.4 Percentage of female students among the total since 1980

Year	Primary	General secondary	Vocational	Higher education
1980	44.6	39.5		23.4
1981	44.0	39.0		24.4
1982	43.7	39.3		26.5
1983	43.7	39.5		26.9
1984	43.8	40.0		28.6
1985	44.8	40.2		30.0
1986	45.1	40.7		25.5
1987	45.4	40.8		33.0
1988	45.6	41.0	43.8	33.4
1989	45.9	41.4	44.5	33.7
1990	46.2	41.9	45.3	33.7
1991	46.5	42.7	45.5	33.4
1992	46.6	43.1	46.1	33.7

Source: *China Education Yearbook*, 1988, 1989, 1990, 1991, and 1992. Advanced Education Press.

Table 12.5 Illiteracy rates of females and males (%)

Year	Female illiteracy Rate	To total illiterates	Male illiteracy Rate	To total illiterates
1982	45.32	69.21	19.15	30.79
1987	38.05	70.12	15.79	29.88
1990	18.50	73.33	6.27	26.67
1995	24.05	72.63	8.98	27.37

Sources:
1982: *Population Census Data*, China Statistics Press, 1983.
1987: *1% Population Sample Survey Data*, China Statistics Press, 1988.
1990: *Population Census Data*, China Statistics Press, 1993.
1995: *1% Population Sample Survey Data*, China Statistics Press, 1997.

there is a significant difference between males and females when the total years of standard education a person has completed are used as an index of his or her educational level (see table 12.6). Using the same index, one can see large differences between male and female education in both urban and rural areas, and the sex differences in education are much bigger in rural than in urban areas (see table 12.7).

Table 12.6 Years of completed education for males and females (%)

	0	1–3	4–6	7–9	10–12	13–16	17+
Female	27.0	14.3	22.3	25.7	8.9	1.6	0.2
Male	9.7	9.6	26.0	36.2	14.8	3.1	0.6

Source: Tao Chunfang, 1993.

Table 12.7 Years of completed education by sex and location (%)

	0	1–3	4–6	7–9	10–12	13–16	17+
Urban female	10.2	7.2	14.9	35.3	25.4	6.6	0.5
Urban male	2.7	3.9	12.5	36.8	29.2	13.0	1.9
Rural female	30.9	16.0	24.3	23.5	4.8	0.4	0.0
Rural male	11.3	11.1	29.1	36.3	11.1	0.9	0.2

Source: Tao Chunfang, 1993.

Table 12.8 The proportion of females who have never been in school by age (%)

	18–19	20–24	25–29	30–34	35–39	40–44	45–49	50–54	55–59	60–64
Urban	1.2	0.3	1.5	2.9	5.0	5.1	10.2	20.7	40.4	50.1
Rural	9.8	15.6	14.3	24.2	33.4	35.8	37.2	63.6	77.2	88.0

Source: Tao Chunfang, 1993.

The number of years of education an individual has received affects his or her ability to progress in society and his or her socio-economic status. Chinese women as a whole are in a disadvantaged position in education compared to men. Educational disadvantage then leads to poor employment opportunities and poor political participation by women.

Age distribution is also very important in analyses of women who have never been in school, because of concern about new female illiteracy, especially in rural China (see table 12.8). A positive correlation between age and the proportion of those never in school is to be expected. As a result of government campaigns to eliminate illiteracy and promote universal primary education since the early fifties, the proportion of people who have never received any education has decreased sharply. However, the difference between urban and rural areas still persists. Although there is also a decrease in rural areas, the proportion of women aged 18–44 who have never received an education is still much higher than that of rural men.

Among illiterate people, some have received education in 'literacy elimination' classes. The criterion for illiteracy elimination is knowing 2,500 of the most frequently used Chinese characters. Using this criterion, the survey found that the percentage of previously illiterate urban women who have now met the standard is 77.9, for urban men it is 91.2. The rural rates are 46.2 per cent and 71.1 per cent for women and men respectively. The age distribution of illiterate women is not even. Due to the effects of traditional culture and social background, there are more illiterate women among the middle and older-age population groups.

There are many kinds of continuing-education programmes that are being run by universities, colleges, and adult vocational schools. Both men and women can take part in these programmes. In addition, there are women's cadre management institutes that provide continuing education specially for females. However, according to the 1990 Women's Status Survey, only 26.1 per cent of women were then taking part in continuing education, which was much lower than the 32.5 per cent of men nationwide. In urban China, the proportions for females and males were 39.1 per cent and 54.6 per cent, whilst in rural areas, the proportions were 22.9 per cent and 27.7 per cent. One can see that females, both urban and rural, are less likely to be receiving continuing education than males, but the difference in proportion between women and men in urban areas is much higher than in rural areas.

Female Employment

The government of the People's Republic of China has been emphasizing women's participation in economic construction and social activities since 1949. Through its laws and policies, the government has actively encouraged women's employment. Hence, the idea that women's liberation depends on deepening participation in social and economic affairs has been embedded in the minds of all Chinese people. There has been substantial equality in labour-force participation between the two sexes, even though sex differentials persist. This holds true particularly when China is compared with other Asian countries (Ogawa and Saito, 1987). In theory, women workers receive the same wages as their male colleagues in the same position. However, the opportunities for job promotion and the occupational structure for men and women are different.

Since the economic reform begun at the end of the 1970s, social and economic growth has provided Chinese women increasingly with opportunities for employment. At the same time, more Chinese women than men, especially in urban areas, have lost their jobs with state-owned enterprises in the reform process, especially in 1990s. The 1990 population census indicated that the overall labour force participation rate of Chinese women aged 15 and above was 72.93 per cent, which is higher than the 70.05 per cent in 1982. The female activity rate increased from 46.6 per cent in 1982

Table 12.9   The overall proportion of women engaged in extra-domestic labour and relative proportion of urban working women, 1978–1990

Year	Overall proportion of women engaged in extra-domestic labour (%)	Relative proportion of urban working women (%)
1978	43.3	32.9
1979	—	34.2
1980	43.4	35.4
1981	43.4	36.0
1982	43.6	36.3
1983	43.6	36.5
1984	43.7	36.4
1985	43.4	36.4
1986	—	36.6
1987	44.5	36.8
1988	—	37.0
1989	—	37.4
1990	44.9	37.4

Sources:
Statistics on Chinese Women, China Statistics Press, 1991, pp. 230, 288; *The Fourth Population Census of China*, China Statistics Press, 1990, Vol. 2, p. 296; *1990 Chinese Labour and Wages Yearbook*, China Statistics Press, 1990, p. 11.

to 53.04 per cent in 1990, which was higher than some developed countries like the United States, Japan and the European Community countries (Zhu Chuzhu, 1991). Table 12.9 shows that during 1978–90, both the overall proportion of women's participation in extra-domestic labour and the proportion of urban working women increased steadily. About 70.82 per cent of Chinese women aged 15 plus were employed in 1995, but marked regional differentials existed (see table 12.10).

The rate of female labour force participation in different age groups has changed in three main respects during the 1990s (see table 12.11). First, the participation rate of young women aged 15–19 years old has decreased, although it is still relatively high. This is certainly related to the increase in educational attainment of young women in high schools and universities. The labour participation rate of younger women was clearly higher than that of developed countries, and even some other developing countries (Tan, 1990). Second, the labour participation rate of women aged 20–44 was very high (over 80 per cent). This not only indicates that marriage and reproduction factors have not greatly affected the labour participation rate of women, but also means that women in China have to play multiple roles in society and family. The activity rate of women aged 50 and over rapidly decreases with increasing age, but it increased from 1982 to 1990, perhaps

Table 12.10 Female economic participation by province (%)

Provinces	Percentage of female population aged 15+ in employment	Provinces	Percentage of female population aged 15+ in employment
Beijing	59.98	Henan	78.49
Tianjin	57.07	Hubei	73.41
Hebei	72.45	Hunan	71.79
Shanxi	56.86	Guangdong	65.03
Inner Mongolia	63.63	Guangxi	76.23
Liaoning	62.28	Hainan	68.41
Jilin	60.32	Sichuan	78.42
Heilongjiang	54.03	Guizhou	77.55
Shanghai	62.02	Yunnan	80.05
Jiangsu	76.10	Tibet	75.64
Zhejiang	59.66	Shaanxi	70.14
Anhui	76.02	Gansu	76.37
Fujian	57.96	Qinghai	71.11
Jiangxi	68.17	Ningxia	73.81
Shandong	73.86	Xinjiang	65.67

Source: *1% Population Sample Survey*, China Statistics Press.

Table 12.11 Female labour force participation rate by age (%)

Age group	Female labour force participation rate in 1982	Female labour force participation rate in 1990
15–19	77.82	68.22
20–4	90.34	89.62
25–9	88.77	90.79
30–4	88.77	90.93
35–9	88.46	91.02
40–4	93.34	88.16
45–9	70.57	81.01
50–4	50.90	61.90
55–9	32.87	44.94
60–4	16.87	27.21
65+	4.73	7.95

Source: Wang Wen (1992), Analysis of the conditions and characteristics of Chinese female employees, *Development of Human Resources in China*, No.4.

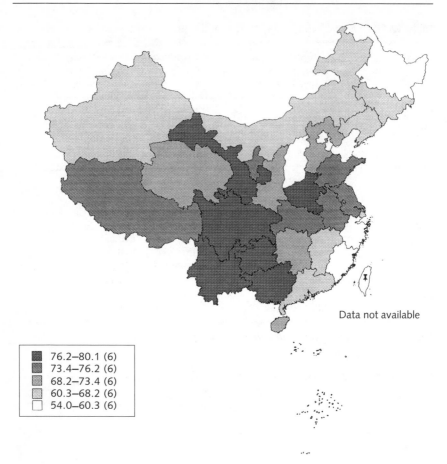

Data not available

76.2–80.1 (6)
73.4–76.2 (6)
68.2–73.4 (6)
60.3–68.2 (6)
54.0–60.3 (6)

Figure 12.1 Percentage of female employed population aged 15+ by province, 1995

Source: As table 12.10

because of the relatively low retirement age of women (50 years old for manual workers, 55 years old for non-manual professionals, and 60 years old for a very few professional and female leaders with high positions). With the relatively high female life expectancy at zero (71.8 years old in urban areas), working women can be expected to live for more than twenty years following retirement. They thus need to find some kind of occupation during this period, while, at the same time, the reform of the employment system means that some women may be allowed to find new jobs after formal retirement.

Compared to male employees, female workers were predominantly concentrated in manual occupations like agriculture (including farming, forestry, animal husbandry and fishery), production and transportation, and commerce, especially in services (51.71 per cent), so that we may speak of the same 'feminization' of services in China as has been seen in other countries. At the same time, men far outnumbered women in professional occupations. This reveals that while women and men may be entitled to equal access to participation in labour, women, in fact, do not have equal opportunities to participate in non-manual occupations. Even within the non-manual occupational sector, female employees are concentrated in the jobs with lower status. They are less likely to work in decision-making positions or be considered for promotion (see table 12.12). In the 1990s, more female labourers have been shifting to the service sectors. This pattern seems to indicate that the equality in employment opportunity provided by the Chinese government has been realized at the cost of inequality in occupational structures and promotion opportunities.

From the occupational composition of the female workforce, one can clearly see (in table 12.13) the high concentration of women workers in farming, forestry, animal husbandry and fishery, and also the relatively great difference and segregation between men and women. In the same table it is also noticeable that there is a very low percentage of women in charge of state organs, parties, non-governmental organizations (NGOs), enterprises and institutions, offices etc. In all kinds of managerial or decision-making positions, the proportion of males is 6.24 times larger than that of females. Even among office workers and related personnel, males outnumber females 2.47 times.

The Chinese Constitution, the Law Protecting the Rights and Interests of Women and other laws and regulations clearly stipulate that the State will protect the rights and interests of female employees, and stresses that

Table 12.12 The proportion of Chinese female employees in different occupations, 1990

Occupation	Proportion (%)
All kinds of professional and technical personnel	45.10
Leaders of the government, parties and NGOs	11.20
Offices and related personnel	26.04
Commerce workers	46.77
Service workers	51.71
Labourers in farming, forestry, animal husbandry and fishery	47.90
Manufacturing and transportation workers and related personnel	35.80
Others not easily classifiable	41.70

Source: Population Census Office under the State Council and Department of Population Statistics, State Statistics Bureau, People's Republic of China *Tabulation on the 1990 Population Census of The People's Republic of China*, China Statistics Press, 1993, Vol. 2, 573–601.

Table 12.13   The composition of working females and males by occupation (%)

Occupation	Female	Male
All kinds of professional and technical personnel	5.35	5.28
Leaders of the government, parties and NGOs	0.45	2.81
Office and related personnel	0.95	2.35
Commerce workers	3.12	2.92
Service workers	2.75	2.11
Labourers in farming, forestry, animal husbandry and fishery	75.26	66.76
Manufacturing and transportation workers and related personnel	12.03	17.72
Others not easily classifiable	0.09	0.05
Total	100.00	100.00

Source: Population Census Office under the State Council and Department of Population Statistics, State Statistics Bureau, People's Republic of China *Tabulation of the 1990 Population Census of the People's Republic of China*, China Statistics Press, 1993.

female and male employees must receive equal salaries and benefits for the same work. Under such a prerequisite, the differences in pay between male and female employees can be seen to be mainly caused by gender segregation in occupations and industry. The distribution of the income levels of women and men in cities and towns is shown in table 12.14.

The most notable features of the levels of pay for female employees and the differences between females and males in cities and towns are the following. First, the average monthly income of females is only 77.45 per cent that of male employees. Second, female labourers are more concentrated in the low-pay groups, 73.7 per cent of women having a monthly income under 200 RMB yuan, while the same proportion for males is only 62.5 per cent, a difference of 11.2 percentage points. More serious is that more than one-quarter of women belong to the lowest-paid group, while less than one-tenth of males are in that group.

The incompatibility between women's productive and reproductive roles has become more serious with the progress of the economic reform in China and the establishment of a market economy in which efficiency rather than equality is becoming the major concern of society. In the last few years a measure of 'optimization of labour composition' has been carried out widely, especially in large state-owned companies. The general purpose of this measure is to improve efficiency and reduce deficits. Economically speaking, this makes sense. As a result of the move, a relatively large proportion of unfit workers have been transferred to other economic activities which require less training and skill or made to stay at home waiting for

Table 12.14 The distribution of monthly income levels for women and men employees in cities and towns (%)

Income (RMB)	Women	Men
0	5.3	1.0
1–49	2.7	0.6
50–99	17.1	8.2
100–49	32.2	26.8
150–99	21.7	26.9
200–49	11.5	16.9
250–99	3.4	7.1
300–49	2.4	5.7
350–99	1.0	2.0
400–99	1.9	3.3
600–99	0.3	0.8
above 800	0.4	0.6
Average monthly income in 1990	149.60	193.15

Note: There are both 0.1 per cent of women and men who do not know their monthly income status.
Source: Calculated according to the data of *The Women's Status Survey in China in 1990* (L. Tan, 1995).

jobs. As the entire labour force faces this challenge, women workers are in a much more vulnerable situation, since they are regarded as more costly and less efficient even by some economists. The number of unemployed women has increased substantially. Women are shocked when they find their formerly secure jobs have now become suddenly insecure, and they are threatened by unemployment more often than their male counterparts.

Discussion and Conclusion

Since the founding of the People's Republic, tremendous social reform has been carried out. Chinese women have achieved historic progress in elevating their status. Nevertheless, the root of gender inequality has, to a large extent, not been removed. Moreover, there is almost no clear-cut gender theory on which relevant policy is based and implemented. Women's issues are frequently simplified as a matter of economic participation. Consequently, government policies regarding women often lack consistency and can be incompatible with other socio-economic policies, which greatly reduces the effectiveness of such efforts to advance women. This is very obvious in the current transitional period. It should also be noted that this situation is not unique to women's issues, but also occurs in other fields such as population control. More integration is needed to combine the task of the advancement of women with the establishment and restructuring of China's market economy, social welfare and security systems, education and health services etc.

The expectation that women's status will automatically and linearly improve along with economic development in general, or with women's increasing participation in economic activities in particular, may not be realistic but very simple-minded. The high rate of women's economic participation has not greatly reduced the gender difference in many aspects of social life. The fundamental social attitude towards women's participation in the labour force, however, has not changed by and large. There have been overt discussions in the Chinese media of the desirability of a 'return to the kitchen' by the Chinese woman worker in the last few years. To ensure the improvement of women's status along with the process of economic growth, and to have a situation in which women's status and economic development may enhance each other positively, requires a great deal of effort.

Another issue is the legal status that women are entitled to have, and the actual status women hold at home and in society. Responding to the Fourth World Conference on Women, The Programme for the Development of Chinese Women (1995–2000) was formally promulgated by the State Council of the People's Republic of China in July 1995. The programme stipulates the tasks and major objectives of the development of Chinese women (All-China Women's Federation, 1995). The National Working Committee for Children and Women under the State Council of China is responsible for the implementation of the programme. In addition, corresponding mechanisms have been set up in each province, autonomous region and municipality under the jurisdiction of the central government. The laws and regulations are there. The question is how can women use these laws to protect themselves against discriminatory treatment and to carry out the purposes women have determined. This becomes even more urgent and important as the society undergoes rapid transition.

As we mentioned early, the old system basically operated through women's work units and places of residence. It was reliant on a strong government administration system with an ideological umbrella and very limited mobility in both geographic and social terms. Economic reform in China has brought about many new forms of employment and great freedom for geographic mobility. The old mechanism through which the equal laws and regulations used to be implemented is under attack. At the same time, the efficiency of women's organizations is declining. Women's Federations which have played a very important role in the struggle for women's rights, do not fully function in many foreign companies, joint-ventures and many newly established private firms. The organizational transformation of relevant government agencies is also under way. To some extent, there is a vacuum. Moreover, China's Women's Federation itself should also modify its working strategies and methodologies to adapt to the changing situation.

On the other hand, women should make full use of the available legal weapons to protect themselves. Survey data shows that, even in big cities like Shanghai, women's awareness of their legal employment rights is low, especially when they are discriminated against (FDWSC, 1999). Many legal service centres or hotlines have been specially established by government agencies and NGOs to provide free legal consultation to women in need. Some of them have worked very successfully in the past. However, most of these centres are urban-oriented. The access of rural women to legal aids is limited. Furthermore, as the legal awareness of the whole society is relatively low, it is much more difficult for women to protect themselves by using legal means.

Up to now poverty in China has been basically a rural phenomenon, although urban poverty has recently become an issue attracting attention. It is commonly believed that women are suffering more from poverty than their male counterparts, although more studies are required to clarify the notion. A few emerging phenomena are noteworthy. Sex ratio at birth has increased continuously since the mid-1980s.[1] This may in general reflect the deterioration of Chinese women's position in society, even though the actual cause of the abnormality is a topic for further exploration. Another issue is the criminal acts such as the abduction, trafficking, and abandonment of women, which often occur in the poor regions of China. The Chinese government has taken very tough measures to curb such crimes. But there is a long way to go in order to eliminate them entirely.

One of the measures of poverty alleviation is to send surplus rural labourers, especially the young, to cities through organized arrangements or the labour market. Incomes generated by those migrant women, in many cases, have already lifted the women and their families out of poverty. Those who migrate from rural to urban areas are commonly those with a better-than-average education. Although their wages and working conditions may be, in some cases, inferior to those of indigenous urban women, they are definitely not marginalized or in absolute poverty. On the contrary, they are obviously economically better off as a result of their migration and their jobs than they were before. This is similar to what Foo and Lim found among women workers in Asia's EPZs (export processing zones) (Foo and Lim, 1989). To some extent, we can say that the temporary rural-urban migration is an enlightened movement that will have great impact on the empowerment of Chinese women.

Gender subordination can be found in many aspects of Chinese social life. However, it may not take the same form as it does in other developing and developed countries. It is important to recognize this diversity while working in a more general framework. The government and scholars should rethink the strategies of pursuing gender equality and should make great efforts to restructure the operational system in order to adapt to the new socio-economic context. Otherwise, China may fall into the trap of allowing women's status to deteriorate during the take-off stage of economic development, as occurred in some of the developing countries.

Note

1.   Details can be seen in ch. 6.

References

All-China Women's Federation (1995), *Programme for the Development of Chinese Women (1995–2000)*. Beijing.
Foo, G. H. C. and Lim, L. Y. C. (1989), Poverty, ideology and women export factory workers in Southeast Asia. In H. Afshar and B. Agarwal (eds) *Women, Poetry and Ideology in Asia*, Macmillan.
Fudan University Women's Study Centre (FDWSC) (1999), *Report on Women's Legal Awareness in Shanghai*. Shanghai: NOVIB Project Report.
Greenhalgh, Susan (1991), International comparison of women's status. In Zhu Chuzhu and Jiang Zhenghua (eds) *Female Population in China*. Henan People's Press (in Chinese).
National Population Sample Survey Office (1997), *1995 National One-per-cent Population Sample Survey*. China Statistics Press.
National Population Sample Survey Office (1997), *1997 National One-per-cent Population Sample Survey*. China Statistics Press.
Ogawa, Naohiro and Yasuhiko, Saito (1987), Male-female differentials in labour force participation in contemporary China. In UN ESCAP, *Women's Economic Participation in Asia and the Pacific*.
Population Census Office under the State Council and Department of Population Statistics, State Statistics Bureau, People's Republic of China (1983), *Tabulation of the 1982 Population Census of the People's Republic of China*. China Statistics Press.
Population Census Office under the State Council and Department of Population Statistics, State Statistics Bureau, People's Republic of China (1993), *Tabulation of the 1990 Population Census of the People's Republic of China*. China Statistics Press.
Tao, Chunfang (1993), *Zhong Guo Fu Nu She Hui Di Wei Gai Guan* (*General Situation of Women's Social Status in China*), China Women's Press.
UNDP (1999), *Human Development Report*.

13   Urbanization

Zhong Fenggan

Studies of urbanization have revealed the unique nature of the process in the Chinese context (Kirkby, 1985). The statistical data, level, mechanism, and areal variations of China's urbanization help reveal this uniqueness.

Process Review, Critiques of Data and Policy Transition

An overview of China's urbanization is given in table 13.1. The years selected for this table were chosen on the basis of the following criteria: (1) being a census year (1953, 1964, 1982, 1990) or one of the years before or after a census (in 1990, the census data are directly used); (2) being one of the years of the Great Leap Forward or the 'adjustment' immediately after it (1958–62); (3) not being one of the stagnant years of the Cultural Revolution (1966–76) (these were omitted); (4) falling between the end of 1970s and the beginning of 1990s (more years were selected from this period to reflect the trend of a relatively rapid increase in the urban population).

Two additional points must also be stated here. First, for most of the period since the founding of the People's Republic in 1949, there are two basic statistics of urban population, the total urban population (including agricultural and non-agricultural population in the urban territory) and the non-agricultural population. Second, for better comparison, adjustments should be made to the official data during the period when inadequate statistical standards were applied (mainly after the middle of 1980s).

In the 40 or more years under our consideration, China's urbanization level (measured by urban population percentage) started from a low of just over 10 per cent and increased to just less than 30 per cent at present. But the annual rate of increase of the urban population is twice that of total population (4.2 per cent vs 1.8 per cent). So, China's level of urbanization was and still is low, but the progress of urbanization has been rapid. Figure 13.2 presents the process of China's urbanization over the past five decades.

Previous studies of the urbanization process in China have highlighted the fluctuating trends in urbanization during the past several decades (Xu Xue-Qiang et al., 1988 and Gao Pei-Yi, 1991). As table 13.1 illustrates,

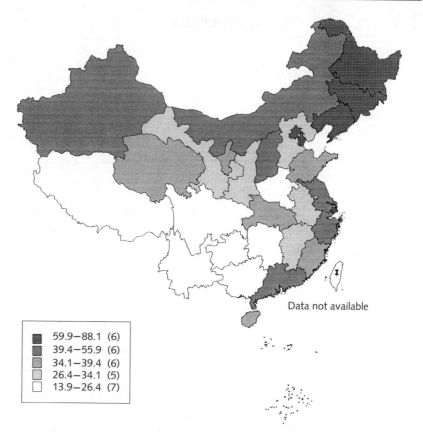

Data not available

59.9−88.1 (6)
39.4−55.9 (6)
34.1−39.4 (6)
26.4−34.1 (5)
13.9−26.4 (7)

Figure 13.1   Total urban population percentage by region, China, 1990
Source: Based upon the data from table 13.2.

we can identify different periods. Before the Great Leap Forward (1958), China's urbanization was 'normal' with an even yearly increase in urban population percentage throughout the whole period. This was because the political campaigns in this period only slightly influenced normal economic development and population migration. During and following the Great Leap Forward period (1958–63), the urbanization process rose and fell sharply (the percentage of urban population went from 16+ per cent to − 20 per cent and back to 16+ per cent again). This can be explained by the Great Leap Forward policy, which encouraged rural-urban migration and an immense influx of people to cities over a short period of time. This rapid increase resulted in the introduction of control methods to curb rural to urban migration. The introduction at this time of a new statistical classification for the urban population (agricultural and non-agricultural)

Table 13.1 Urban population and its percentage of the total population of China, 1949–1996 (selected years)

	Urban population (10,000)			Urban population percentage (%)		
	Official	Adjusted	Non-agricultural	Official	Adjusted	Non-agricultural
1952	7,163			12.46		
1953	7,826			13.31		
1954	8,249			13.69		
1957	9,949			15.39		
1958	17,021			16.25		
1959	12,371			18.41		
1960	13,073			19.75		
1961	12,707		10,602.7	19.29	16.1	
1962	11,659		9,818.8	17.33	14.6	
1963	11,646		10,007.1	16.84	14.5	
1964	12,950		9,885.4	18.37	14.0	
1965	13,045		10,169.9	17.89	14.0	
1970	(14,424)			(17.38)		
1975	(16,030)			(17.34)		
1977	16,669		11,955.9	17.55	12.6	
1978	17,245		12,444.2	17.92	12.9	
1979	18,495		13,312.3	19.96	13.6	
1980	(19,140)			(19.36)		
1981	20,171	18,033	14,320.3	20.18	18.30	14.3
1982	21,154	18,562	14,715.0	20.83	18.26	14.5
1983	24,150	19,443	15,234.2	23.50	18.97	14.8
1984	33,136	21,492	16,689.4	31.90	20.77	16.1
1985	38,446 (25,094)	23,311	17,970.8	36.60 (23.71)	22.30	17.1
1986	44,103 (26,366)	24,295	19,191.1	41.40 (24.52)	22.98	17.1
1987	(27,674)			(25.32)		
1988	54,369 (28,661)	27,059	20,087.1	49.60 (25.81)	24.83	18.3
11989	(29,540)			(26.21)		
11990	29,651 (30,191)			(26.41)	26.20	
1991	(30,543)			(26.37)		
1992	71,234 (32,372)	31,924	23,412.0	61.64 (27.63)	27.70	20.25
1993	(33,351)			(28.14)		
1994	(34,301)			(28.62)		
1995	(35,174)			(29.04)		
1996	(35,950)			(29.37)		

Note: Data in ( ) are published by SSB in *China Statistical Yearbook, 1997.*
Source: Data complied from publications by China State Statistic Bureau, including *Statistical Yearbook of China*, 1983, 1989 and 1993 versions; *Collection of Population Statistics of PRC (1949–85); Population Statistics of Municipalities and Counties 1988, 1989; Fourth Population Census.* Adjusted data are from *Study of Regional Development: Problems of Chin's Urbanization* edited by Wang Sijun, published by Higher Education Press, 1996.

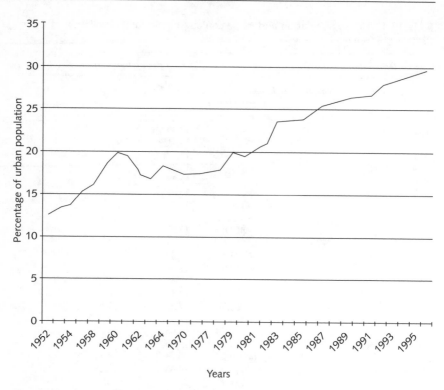

Figure 13.2 Process of urbanization in China, selected years

reflects this awareness. This disaggregated urban population classification, begun in 1963–4, showed that the authorities were concerned about rural migration to the urban areas, drawn by the faster economic development occurring in urban regions.

From 1965 until China's adoption of reform policies and its opening to the outside world (1978), two factors made the urbanization process apparently stagnant. During this period the urban population remained at 17 per cent. The first was simultaneous rural-urban and urban–rural migrations, the latter outweighing the former (Kirkby, 1985). The other was a disguised urbanization process caused by relatively faster demographic growth in urban areas (Zhong and Clarke, 1985). Since the winter of 1978, the reform and opening era have given China's economy and urbanization relatively lasting impetus. Both the adjusted and the non-agricultural data show a steady increase in urbanization level (from 18+ per cent to 27+ per cent and 14+ per cent to 18+ per cent, respectively). Yet, it should be noted that the information available for the period 1984 onward sometimes causes confusion as China's statistical system changed its criterion for urban population.[1] Later, the State Statistical Bureau of China realized the problem, and made its own adjustments, which made much more sense. In table 13.1, both indices using different criterion are presented to demonstrate the necessity of data adjustment in China's urbanization in and after 1980s (Wang Sijun et al., 1996).

The review above impels us to say something more about the mechanism of China's urbanization. For the first three decades after the founding of the People's Republic, China had a centrally planned economy. The urbanization process was closely conditioned by explicit and implicit population redistribution policies. One of the objectives of the planned economy was to redress rural to urban migration resulting from the movement of country dwellers to cities in search of better living conditions and employment opportunities. The government's basic explicit population redistribution policies were intended to influence (restrain) rural–urban migration. The fundamental mechanisms to ensure this were the 'permanent residence registration system' and the 'rationing system'. On the basis of these policies, other measures such as restricting the influx of rural population and rusticating urban residents to rural areas worked more 'actively' to control the increase of urban population.

Related to this was China's implicit population redistribution policy to assist a planned increase of urban population. The overall economic policy for urban areas was firstly to 'turn the consumer city into the producer city'. This policy neglected the development of the tertiary sector for a long time and kept unemployment rates higher than they might have otherwise been in urban areas. The 'Two Legs' development strategy was another implicit population redistribution policy that had broader implications. The main idea of this policy was to strive to obtain the balanced development of agriculture and industry, of heavy and light industry, of coastal and inland industry, and of industry in large, medium and small cities. Leaving aside the internal ideological basis of the 'Two Legs' strategy it helped urban population growth in inland and less inhabited areas and led to a rapid increase in the urban population in medium and small cities. This lessened the geographical concentration of the urbanization process, and led to different urban growth processes in different areas over time.

All these policies, either explicit or implicit, were interwoven and implemented simultaneously during various political campaigns – the Great Leap Forward, the Socialist Education Movement, 'Young Graduates Going to Countryside', and the Cultural Revolution. The major outcome of these policies was a restriction of urban population growth. The long-term trend contrasts with urbanization processes elsewhere as the simultaneous rural to urban and urban to rural migration favoured the latter.

The two-way migrations between 1965 and 1976 resulted in a net migration of more than 6 million from urban to rural areas. During the Cultural Revolution period the urban–rural migration was mainly characterized by the fact that 20 million urban students were sent out into the countryside all over China. This resulted in another counter-migration in the later 1970s as those students returned to their urban origins. We also see a surge of 30 million from urban to rural in the 1961–64 period. But there were two occasions when rural-urban migration surges prevailed throughout the country: the 1958–9 surge of 20 million peasants and the 1977–82 migration of 33 million former urban residents (Kirkby, 1985). Yet these policies and their effects and their influences have been undergoing a transition since the winter of 1978 when 'the era of reform and openness' began. The policy changes were most noticeable in changing the balanced economic development programme and concentrating on economic development where the efficiency criterion was present.

Though the permanent residence registration system (the urban–rural dual system) cannot be easily abolished – a policy to abolish it has, nevertheless, been considered recently – a specific policy to allow peasants to migrate to rural market towns or even other urban areas for activities other than agriculture since the winter of 1984 has meant the relaxation of the control of rural-urban migration. These migrating populations were considered by official statistical records as 'non-agricultural population'. This is the main change in explicit population redistribution policies.

There are two important changes to the implicit population redistribution policies. The first one is that instead of the policy of 'turning the consumer city into the producer city', which neglected the development of the tertiary sector, a policy for achieving balanced development of primary, secondary and tertiary sectors in urban areas has been implemented since 1985 (Zhong et al., 1985). The second results from increasing domestic and foreign investment in more developed areas (mainly cities and towns in coastal areas) which draws agricultural labourers to these areas. These labourers have no 'permanent residence registration' in urban areas, but they are regarded as urban population in official statistics with a 'temporary residence registration'.

The above review of population redistribution policies gives greater depth of understanding to the urbanization dynamic as numerically presented in table 13.1. The changing policy contexts require us to notice a kind of urbanization taking place in some populous rural areas because of rapid economic development promoted by the peasants themselves, but not included in official statistics.

Following other published research, urbanization measured by official data is referred to as 'apparent urbanization' and urbanization in rural areas not indicated in official data is termed 'latent urbanization'. 'Latent urbanization' is a kind of urbanization where the population already possess some or more urban characteristics in occupation and life style but does not have residence-transference status (Wang et al., 1996). However, no matter what kinds of measures are used, China has followed what the World Bank called in 1982 a road of 'industrialization without urbanization'. The level of urbanization in China is not only far below that of developed countries, it is also lower than the average for developing countries.

We can describe the mechanisms forming China's urbanization pattern as follows: Generally, we may regard the urbanization process as an effect-accumulation of explicit and implicit population redistribution policies, as discussed above. Determinants of change include the following factors:

1. The land available is inadequate for the population and the labour force in rural areas and this ratio has been getting worse since the founding of the People's Republic because of continuing demographic increase and improvements in productivity.
2. The emphasis on heavy industry to the neglect of developing the tertiary sector has seriously reduced the labour-absorbing capacity of industry.
3. The effective implementation of the policy to restrict migration to urban areas kept the labour force and population in rural areas.
4. In the process of policy transition, the causes and effects of government investment are not clearly understood, but private investment (from home or abroad) in some rural areas has produced marked effects in occupational transition and improvements in the standard of living (Wang et al., 1996).

It is difficult to estimate the total latent urban population, especially when rural labourers can choose to live in rural areas but are employed in urban areas. Yet, based on field investigation, some research results hold that, for the whole country, there are 131.9 million people who would classify as latent urban population. This accounts for 11.6 per cent of the total population, or 14.7 per cent of the total rural population, of the whole country. This latent urban population is only a little less than half (46 per cent) of the apparent urban population (Wang et al., 1996).

Regional Variations

China is a large and populous country with a diverse natural environment and varying geographical condition and levels of socio-economic development. Naturally, regional variations in urbanization need to be discussed in some detail. Firstly, it should be recognized that the present regional variation in China's urbanization is an outcome of the accumulation of past development. Since the founding of the People's Republic this development has gone through a policy transition: from the 'space-balanced development policy' implemented before 1978 to the 'disequilibrium and polarization development policy' implemented since 1978. In essence, this transition has close ties with the transition from explicit to implicit population redistribution policies mentioned above. While it is obvious that the former policy would reduce the areal variations in urbanization inherited from old China, the latter policy might lead to an expansion of the existing areal variations. Moreover, while this latter policy is being implemented, the adoption of a policy of reform and openness is drawing more and more investment from home and abroad promoting urbanization in the areas where investment occurs.

There have been concerns about variations in apparent urbanization since the beginning of the 1980s. Research highlights the fact that the variations depend on the differential economic development under certain space-development policies and on the basis of different natural endowments. Comparative variations in latent urbanization were the focus of academic attention only very recently (Wang et al., 1996). Recent research suggests that variations in latent urbanization arise from differential development, especially from the differential development of rural areas. This is influenced by private investment or by the expansion in production occurring in areas of apparent urbanization during the reform era.

An analysis of these areal variations using provinces (or autonomous regions) as units is found in table 13.2. This table lists the apparent and latent urbanization levels (in population percentages), the economic development level (in per capita GNP), industry structure (light and heavy industry) and population density of every province. As indicated by other researches, apparent urbanization levels in coastal areas are higher than in middle (inland) and west (border) areas, and also higher in the north than in the middle and southern regions (see table 13.2, column 1). But, as far as the latent urbanization level is concerned, things are quite different.

If we take China's coastline and the Yangtze River as roughly forming a letter T lying on its side (—|), the highly urbanized provinces, by Chinese standards, can be found along the lines of this —|. In addition, the level of urbanization decreases as one moves from east to west, and also as one moves north and south from the centre (see table 13.2 column 4).

The five regional groupings used in table 13.2 are: I the highly urbanized coastal region; II the latently urbanized coastal region; III the highly urbanized region in the north; IV the central region with a medium level of urbanization; V the region with a low level of urbanization in the south (see also figure 13.1).

The four provincial units included in Type I are those with high level of apparent or latent urbanization. In the 'pre-opening' era, this populous coastal region both inherited high historical development levels, and also benefited from the industrial development policies of 'Two Legs' and 'War Preparedness', which enhanced its urbanization process. In the opening era, urbanization in the region further benefited from the 'disequilibrium and polarization development policy'. Industrial and urban development levels remained relatively high and, in rural areas, private investment from home and abroad induced a marked latent urbanization process. It is obvious that high productivity, especially in the heavy industry sector, and the high urbanization level of this region reflect the economy's strength and development potential.

Table 13.2 China's urbanization by types of region, 1990

Type	Province	% of apparent urban pop.	% of latent urban pop.	% of total urban pop.	% of latent urban pop. to rural pop.	Per capita GNP (yuan/pers on)	Ratio of light to heavy industries	Population density (person/km²)
I	Liaoning	52.60	10.47	63.07	21.32	2,432.29	0.48	270
	Beijing	73.66	14.35	88.01	39.77	4,610.68	0.79	644
	Tianjin	69.65	15.24	84.89	34.27	3,397.17	1.04	777
	Hebei	18.07	14.97	33.04	17.69	1,331.31	0.94	325
II	Shandong	22.29	15.49	37.78	18.65	1,568.50	1.03	539
	Jiangsu	24.52	21.95	46.47	27.71	1,942.35	1.21	645
	Shanghai	27.82	15.75	83.57	50.40	5,569.71	1.08	2,118
	Zhejiang	21.72	21.05	42.77	24.71	2,007.70	1.87	407
	Fujian	21.66	14.04	35.70	17.11	1,533.88	1.63	248
	Guangdong	31.46	22.91	54.37	27.75	2,319.32	2.03	353
	Hainan	25.61	8.52	34.13	12.62	1,433.03		193
III	Heilongjiang	51.70	4.94	56.34	9.38	1,791.53	0.51	78
	Jilin	51.13	4.78	55.91	8.10	1,586.39	0.72	132
	Inner Mongolia	38.06	5.29	43.35	8.09	1,325.10	0.72	18
	Ningxia	29.08	5.68	34.76	7.69	1,298.94	0.39	90
	Xinjiang	36.78	2.66	39.44	5.01	1,647.35	0.97	9
	Qinghai	30.41	5.33	35.74	7.87	1,479.46	0.41	6
	Shanxi	29.15	16.92	46.07	21.32	1,373.72	0.34	184
	Shaanxi	24.11	8.96	33.07	11.70	1,130.43	0.74	160
	Gansu	20.78	9.10	29.88	11.08	1,039.42	0.40	49
IV	Henan	15.87	10.07	25.94	11.82	1,035.66	0.84	512
	Hubei	27.15	9.53	36.68	12.62	1,457.14	0.90	290
	Hunan	18.84	7.49	26.38	8.79	1,146.61	0.80	286
	Anhui	18.24	9.98	28.22	11.83	1,068.79	1.08	404
	Jiangxi	21.74	9.96	31.70	12.47	1,094.88	0.81	226
	Sichuan	17.90	7.76	25.66	8.99	1,061.35	0.86	188
V	Guizhou	15.47	4.84	20.31	5.58	778.79	0.74	184
	Yunnan	14.72	5.08	19.80	5.87	1,061.35	1.11	94
	Tibet	10.89	3.02	13.91	3.55	1,101.35	0.43	1.8
	Guangxi	15.66	5.90	21.56	6.77	921.92	1.24	178

Sources: Wang Sijun et al. (1996), *Study of Regional Development: Problems of China's Urbanization*; Major Figures of the Fourth National Population Census of China, 1991.

In contrast to the provinces of Type I, the provinces of Type II did not benefit much from the 'space-balanced development policy' of the pre-opening era. Rates of industrialization and urbanization were relatively low, with only a few exceptions (as in and around Shanghai, for example). On the other hand, under the explicit population redistribution policies intended to control rural-urban migration, the potential urbanization growth rate was not realized. So, the low level but relatively high quality of urbanization in this region was significant in the pre-opening era. However, during the opening era's policy transition, coastal areas witnessed a pronounced latent urbanization process as a result of private investment. The development of secondary and tertiary sectors in both urban and rural areas in this region drew large numbers of migrants (floating population) from other regions of the country. As latent urbanization becomes apparent, this region will become a region with even higher urbanization levels in the not too distant future.

Provinces and autonomous regions of Type III benefited greatly from the 'space-balanced development' policy in industrial and urban development before the policy transition. But the natural endowments of most of the nine units in this region were not conducive to development. Thus the general productivity of the area was limited in both rural and urban areas.

Also, most parts of this region have low population densities. In contrast, industrial and urban development promoted by the space-balanced development policy drew population chiefly to cities. Thus the combination of low population density and population concentration in cities resulted in high urban populations. Rural areas with their disadvantageous geographical position and low level of development were unable to attract private investment. The latent urbanization process in this region is less marked than that in Type I and Type II even after the policy transition. Socioeconomic development policy should focus on equalizing the urban development process and improving conditions in rural areas.

The six provinces of Type IV are inland provinces, occupying a section of the middle part of China. Their characteristics are 'medium'; they have average natural endowments, a legacy of mediocre development performance, and average beneficial results stemming from the space-balanced development policy. Productivity in this region is average; the apparent and latent urbanization levels are medium too. In the near future, socioeconomic development policy should promote construction in urban and rural areas on the basis of making full use of the development diffusion from the coastal provinces and in order to build up the productivity of the local labour force. This will produce a balanced urbanization process.

The last type, Type V, consists of four provinces or autonomous regions which are peripheral areas, either located inland, along the border or on the coast far from development centres or poles. Though the natural endowments may not be absolutely poor in this region, the development indicators are the lowest in China. These areas have benefited minimally from the space-balanced development policy. But it is not unrealistic to say that investment in their good natural-resource base and to increase productivity could improve levels of development and increase the rate of urbanization in them.

Conclusion

First, productivity and urbanization levels inherited from old China were low. Though various methods were tried to improve productivity and urban development, China's urbanization process was uneven in the pre-opening era. The policy transition that took place in the winter of 1978 had a normalizing effect on urbanization. Different 'statistical standards' used in official data have created problems in accurately interpreting urbanization processes, but the analysis above identifies the problems and offers a nuanced interpretation. Unlike the apparent urbanization expressed in official data, latent urbanization is not reflected in official statistics. Understanding latent urbanization is important to China's current urbanization trends.

Second, areal variations of China's urbanization show latent as well as apparent urbanization. As far as apparent urbanization is concerned, the disequilibrium pattern was at one time reduced when the space-balanced development policy was implemented. But the disequilibrium and polarization development policy, implemented during the 'opening era' led to a new round of unequal patterns of urban development. For latent urbanization, new development policy after 1978 has promoted differential development among the various regions, with the populous coastal and provinces bordering the Yangtze River developing more rapidly. Each of the five regional types is marked by its particular apparent and latent urbanization characteristics.

Third, in the foreseeable future, the localization and concentration of economic growth and efficiency-priority will still be the essential determinant to socio-economic development in China. Given this, the disequilibrium and polarization development policy will continue. This does not imply that the policy has abandoned criteria of social justice. The more developed regions will in turn assist and contribute to the development of the less developed regions. Thus, at a certain stage, the gap in development and urbanization levels between more developed and less developed regions will become larger, but in the long run, along with the development of the common cause, this gap will be gradually reduced. What should be done in every region, regardless of its level of development and urbanization, is to make full use of internal and external conditions to promote the economy and organize urban development.

Note

1. The essence of this change was that the whole of the agricultural population living within the city boundaries was included in the urban statistics.

References

Gao, Peiyi (1991), *Comparative study of Urbanization between China and Foreign Countries*. Tianjin: Nankai University Press.

Kirkby, R. J. R. (1985), *Urbanization in China: Town and Country in a Developing Economy, 1949–2000*. Croom Helm.

Wang, Sijun (ed.) (1996), *Study of Regional Development: Problems of China's Urbanization*. Beijing: Higher Education Press.

Xu, Xueqiang and Jianru, Zhu (1988), *Modern urban Geography*. Beijing: Publishing House of China's Architectural Industry.

Zhong, Fenggan and Clarke, John (1985), Urban population growth in China: its characteristics and areal variations. *CRU Working paper 26*, Department of Geography, University of Durham, England.

14   The Floating Population and Internal Migration in China

Sun Changmin

China is currently at a historical turning point driven by economic and industrial policy that has brought comprehensive development both economic and social. However, as income levels remain significantly different between urban and rural areas, and between different regions, the scale and influence of the 'floating population' (temporary migrants) have increased dramatically since the second half of the 1980s. According to uncompleted estimates, the total number of the floating population in China was about 50 million in 1989 (Ai, 1989). In 1991 and 1993, it reached 65 million and 80 million respectively (Li, 1991). If these estimates are reliable, the current number of the floating population can be predicted as about 100 million, considering the high economic growth in China since 1992.

A number of surveys suggest that the direction of population movement is from inland to coastal areas, from less developed to developed areas, and from rural to urban areas. The composition of the floating population has presented a series of changes, becoming geographically wider spread, younger, lower educated, and staying away longer from the original place of residence. The large-scale floating population and migration can be seen to have had various and far-reaching effects on the social structure and development process in China. It influences sustainable development in agriculture directly, as well as posing challenges to the urban infrastructure and administration. From these factors stem a range of issues.

Definition and Classification of Floating Population

In China two basic terms are widely used, namely, 'floating population' and 'migrant'. The definition of migrant was given by the National Population Census. The notion of floating population, however, has different meanings under different circumstances.

Definitions and research methodology relating to the floating population

The Chinese notion of 'floating population' refers to those living outside their places of household registration. The definition of floating population in China is different from that of migrants usually used in population studies. It excludes migrants who have household registration at their destinations, regardless of the duration of their mobility. On the other hand, it includes those who have no household registration at their places of destination, no matter how long they have stayed there.

Apart from this general definition, there are various explanatory ones established for different research purposes. One research project, which is jointly undertaken by the Institute of Rural Development of the Chinese Academy of Social Sciences and the Rural Department of the National Statistics Bureau, classifies the floating population into two categories. One consists of the urban population moving between cities and towns, the other of the rural population moving to urban areas. Their focus is on the second category. According to their research, the second category of the floating population can be divided into those as 'floating' for employment purposes and those for non-employment purposes. Generally speaking, the former move from rural to urban areas in spring, and return from urban to rural areas in winter. This pattern of population movement is labelled 'semi-pendulum floating population' (Research Group of Rural Annual Analysis, 1993). In another study of rural labour force movement, the out-moving labour force is defined as those who move outside their residential townships, whether within or beyond the relevant counties and provinces (Zhao, 1994).

The Economics Institute of the Ministry of Urban Construction has drawn up five definitions of the floating population. These five definitions are defined in terms of, respectively, administration, development economics, demography, demographic economics, and demographic geography. In their research project undertaken in 1988 and 1989, the floating population was defined as 'those who have no household registration and are engaged in all kinds of activities in the city'. They focus on the impact of floating population on large cities (Li and Hu, 1991).

To sum up, it is clearly the rural labour force migrating outwards for employment purposes that is the main concern. However, according to the specific stresses of the research, the definitions and classifications of the floating population differ.

Definition of migration

Indicators of migration were not included in the population census in China until the fourth national population census in 1990. According to the definition of the census, those 'whose places of residence changed within the period from 1 July 1985 to 1 July 1990, and changed to new places of residence outside counties or cities' are defined as migrants. The migrants include those migrating within and beyond their provinces. In-migrants are defined as those who have stayed in a place without household registration for more than one year, or who have stayed in a place less than one year but have left their original place of residence, where they have their household registration, for more than one year. Out-migrants only include 'those who have left this county or city for more than one year' (You, 1992; Materials of China's Population Census in 1990, 1993).

Population migration in China can be classified into two categories. One is permanent migration involving change of migrants' places of residence coupled with change of their household registration. The other category is temporary migration, change of migrants' places of residence without change of household registration. The latter and the pendulum-pattern migrants are called the floating population (Gong, 1991: 4). According to the rules of China's household registration administration, population migration occurs when people migrate from their original place of residence to a destination and change their household registration.

The Characteristics of Internal Migration and the Floating Population

Size of floating population

Since the 1980s China's strict household registration system has gradually been relaxed with the shift from a planned economy to a market economy. At the same time, the reform of rural economy and administration has provided more possibilities and flexibility to farmers, in terms of their decision-making on employment and their interest preference at the micro level. The size of the floating population has increased continually.

The results of the fifth investigation of the floating population in Shanghai, conducted in December 1993, showed that the total in-floating population amounted to 2.81 million, of whom 2.51 million came from other provinces and municipalities, Hong Kong, Macao, Taiwan and foreign countries. The total out-floating population amounted to 470,000, of whom 170,000 moved to other provinces and municipalities, Hong Kong, Macao, Taiwan and foreign countries. Another 0.5 million were transients. Hence, the total size of the floating population in Shanghai, including the in-floating and out-floating populations and transients, came to 3.31 million. Compared with a previous survey on 20 October 1988, which gave the total as 1.93 million, the floating population had increased by 1.38 million, and the annual increase rate was 11.47 per cent. At the same time, Beijing, Guangzhou and other regions showed the same trends.

In November 1994 a survey on floating population was conducted. The result showed that the total floating population had reached 3.30 million, of whom 2.83 million came from other provinces and municipalities, 44,000 came from abroad, and 0.42 million were transients. The total floating population amounted to one third of the 10.63 million total resident population in Beijing. It was 3 million more than in 1980, and had increased 15-fold (Ji, 1995).

According to the *Xinming Wanbao* (Xinming Evening News, Shanghai) on 20 February 1994, a new record was set when large numbers of farm workers went to Guangzhou by train immediately after the Spring Festival. On 18 February, about 0.2 million transients arrived in Guangzhou, 80 per cent of whom were farmer workers. As early as in 1987, the total floating population reached over 1 million. It increased to 1.3 million in 1988. The total floating population is approaching 5 million in the cities and towns of the Zhujiang delta region centred on Guangzhou.

In Fujian province the floating population had reached 3.09 million according to a provincial survey undertaken on 15 June 1994. The registered temporary resident population was about 1.65 million, a 190 per cent increase from the 0.58 million of 1989. In some neighbourhoods of the large cities such as Fuzhou, Xiamen and Quanzhou, the temporary resident population was several times larger than the local permanent resident population.

In Sichuan province, according to uncompleted statistics, the total provincial labour force was about 72 million in 1994, of whom 52 million belonged to the rural labour force. Twelve and half million of them shifted to the non-agricultural sector. About 8.79 million left their home towns for work and business, of whom 6 million went to other provinces and municipalities.

According to some estimates, the national size of the floating population increased dramatically in the 1980s. The floating population was about 30 million in 1982. It increased to 40 million in 1985 and reached 60 million in 1987, double the 1982 level. The annual average increase in the number of the floating population was about 10 million for the period.

If the increase of the floating population from 1984 to1988 could be classified as the first peak of high growth, the second peak of high growth started from 1992. The total number of the floating population reached 65 million in 1991 (Li, 1991), and increased to 80 million in 1993. In 1998 it was estimated as about 100 million in total.

Classified in terms of length of stay outside their hometowns, the size and increase rate of the floating population are even more significant. According to the population census of 1982, 6.57 million people among the floating population stayed outside their home towns for over one year. The number increased to 21.35 million in 1990, 3.25 times the number in 1982. It made up 1.88 per cent of the total population in China (Zhang, 1993).

Summing up, the existing estimates of the floating population in China are by no means either systematic or precise owing to the various definitions and statistical methods used. However, recent efforts made by researchers have revealed the significant growth of the floating population. The size of the floating population increased in the early period of the 1990s to more than 10 times what it was in the decade before. Research also suggests that the rapid growth of the floating population is correlated to macro socio-economic development, such as the transformation of the social structure, change of land resource conditions, and variations in the implementation of population policy.

Demographic characteristics of the floating population

Currently, the major direction of movement for the floating population is towards economically developed cities, the industrialized coastal regions, and higher-income areas. The cause of this flow is obviously the imbalance of socio-economic development between regions. Significant variations in production conditions and living standards exist between city and country-side, and between the eastern coastal region and the western inland region. The existence of a large surplus rural labour force causes population outflow and the labour demand created by the rapid economic develop-ment in eastern coastal regions draws population in. The dramatic increase in the floating population today is mainly the result of the function of these push-pull factors.

According to surveys, the floating population in large cities, such as Beijing and Shanghai, presents certain demographic characteristics.

(1) Before the 1980s, social activities, such as visiting relatives or friends, were the main cause of the presence of floating populations in cities. The percentage was above 60 per cent. In the late 1980s, however, economic activity took over as the main cause of the floating population. This resulted in a change of the sex composition of the floating population. According to the fourth national census data, among the floating population, males accounts for 55 per cent and females account for 45 per cent. The sex ratio is about 122, which is 15 percentage points higher than the 106.6 for the total population. The higher sex ratio of the floating population is related to their economic activities at their destinations. Different jobs require dif-ferently sex-orientated labour forces. A survey conducted in November 1994 in Beijing found that males accounted for 63.5 per cent and females account for 36.5 per cent of the floating population. The sex ratio was 174. Classified by economic sectors, the percentage of males was 98.4 per cent in the construction sector, and about 58 per cent in various small-scale busi-nesses involved in selling. The percentage of female was 63.6 per cent in household services (Ji et al., 1995). Another survey undertaken in Sichuan province in 1994 suggests that 73.3 per cent of the 5 million floating popu-lation are engaged in economic activities. Among them, the male labour force is dominant in the construction sector, accounting for 95 per cent. The female labour force is dominant in household services, accounting for 83.7 per cent (Fazhi Ribao, 1994). These findings are also corroborated by other research.

(2) The age composition of the floating population is younger and relatively concentrated. According to a survey in the four cities of Chendu,

Harbin, Anshan and Jiling, the percentage of the floating population under 35 years old was 70 per cent. Those who float from rural areas are even younger (Li, 1991). The floating population is concentrated in the 20–35 age group. In Beijing the number of 15–45 year-olds makes up 81.8 per cent of the total in-floating population. Among them, those aged 15–29 account for 55.3 per cent, those aged 30–45 for 26.5 per cent. In addition, the age group of 0–14 years accounts to 9.9 per cent of the total in-floating population, and only 2.2 per cent consists of those over 60 years old (Liu, 1996). In Shanghai the 20–24 age group is the biggest among the in-floating population. It makes up 22.60 per cent of the total. The age groups 25–9, 30–4, and 15–19 years account for 19.36 per cent, 11.70 per cent and 10.06 per cent, respectively. These four age groups together make up 63.72 per cent of the total in-floating population in Shanghai (Zuo et al., 1995). With the increase in the absolute numbers of the floating population, the age distribution among them becomes much younger and more concentrated. Before the 1980s, the elderly and children amounted to 30 per cent of the total floating population. By the mid- to late 1990s their share declined to about 20 per cent. The proportion of those of working age (15–49 years) accounts for 80 per cent. This can be explained in terms of the exploring spirit of this age group and their lesser burden of family responsibility.

(3) The geographic distribution of the floating population is wider but the direction of movement remains steady. Since the late 1980s nearly every province, municipality and autonomous region has had some degree of floating population, from large cities to medium and small towns, from inland areas to the coastal areas, and from the areas of concentrated Han nationality to those where minorities predominate. The floating population has flowed to every place where the conditions and opportunities for earning more money exist. Those who belong to the floating population often rely on a very strong social network that sharpens the relatively steady direction of movement.

(4) The educational level of the floating population is lower but they show a strong ability to adapt to new circumstances. Several surveys have shown that the educational level of the floating population is low. In Shanghai, for instance, the illiterate make up 9.4 per cent of the total floating population. And only 24.0 per cent and 47.6 per cent have had primary school and junior high school education respectively. Putting them together, the proportion of those below the junior high school account for 81 per cent of the total floating population (Li, 1991). In Beijing, farming was the original occupation of 78.8 per cent of the total in-floating population. The proportion of those below junior high school level education was found to be 85.1 per cent of the total. However, the disadvantage of these lower educational levels is somewhat compensated for by strong capabilities to adapt to new circumstances. The floating population is seen as hard-working and willing

to do all kinds of jobs regardless of working conditions, and as having a strong will to survive and overcome the obstacles of big city life. Though their own lives are hard, these migrants worry most about the education of their children. This has attracted attention from society and the urban authorities.

(5) The members of the floating population tend to stay away from their places of residence longer, and seasonal migration is typical. According to surveys, the average time for those among the floating population to stay away from their original homes is about 195 days. Among them, those who are engaged in household service stay for 307 days, construction workers for 297 days, other employed labourers 236 days, vendors 225 days, and repair workers 207 days. These five categories together make up more than half of the floating population (Li, 1991). In Beijing 16.6 per cent of the floating population stay more than one day but less than one month. 10.8 per cent stay from one month to three months, and 9.3 per cent from three months to six months. Those who stay over six months account for 63.3 per cent of the total floating population (Zhang, 1996). Therefore, most of year, these people stay outside their original places of residence. They return to their home towns only during the holidays, especially during the Spring Festival. Although city life is never as good as in their dreams, most of the floating population are unwilling to go back to the countryside.

The existence of large amount of surplus rural labour is a precondition for the development of cities and urbanization. Under certain circumstances the in-floating labour force brings many benefits to destination centres. However, when the number of the floating population exceeds the capacities of these cities, many problems may arise.

The Impact of Socio-economic Development on Population Migration and Mobility

Recent socio-economic reform and development in China is challenging the old social structure of the planned economy system. Government policy has been, generally speaking, in favour of rural-urban migration. For example, in 1983 the government allowed temporary migration from villages to townships provided the migrants retained their village household registration. In 1984 another policy allowed the permanent settlement of peasants in small towns and cities, though migrants had to provide their own supply of food grain.

The historically imbalanced development between urban and rural areas, and between different regions, is widening in the process of reform. The implementation of the household responsibility system in rural areas has raised agricultural productivity, but has also created a large surplus rural labour force. Together with the huge size of the existing rural population, greater pressure is being exerted on land resources than ever before. In addition, unbalanced fertility levels between regions also cause pressure on employment. Urban places became more tolerant to, and dependent on, the peasant workers' employment in construction, traditional services, and other physically demanding and hazardous jobs that are usually refused by urban workers. In most cities migrants can legalize their temporary urban residence by applying for a 'certificate of temporary residence' and 'work permit'. According to the new regulations in Shanghai and some other cities, 'blue-seal household registration' (*lanyin hukou*) status can be selectively granted to migrants who have made at least a certain amount of investment in either productive projects or housing, or those who have been employed by a local firm for more than three years (Wang, Le and Zuo et al., 1996).

The transformation of the social structure and population mobility

The development of Chinese society in the past decades has been shaped by basic social structures with Chinese characteristics. With the deepening reforms beginning from the early 1980s, the regime of rural people's communes was abolished and replaced by the county as the basic administrative unit. Under the household responsibility system and with the development of rural industries, individuals have been given greater flexibility. For example, farmers have more freedom to arrange their working times and production behaviour. The rural labour force has been released to some extent from the land as agricultural labour productivity has been raised.

At the same time, the rapid growth of township enterprises has played an important role in contributing to the national economy, and in absorbing the 110 million surplus labourers who have shifted from agriculture. The total income of township enterprises accounts for about 50 per cent of the rural economy and one third of the total national industrial output value. However, township enterprises have been challenged by new circumstances. First, the implementation of management contracting systems in township enterprises has resulted in a situation where township enterprises no longer have the responsibility for absorbing surplus rural labour. Secondly, because of market competition, the capacities of township enterprises to absorb surplus labour are becoming restrained.

Surplus rural labour has also moved to cities and towns, as well as to coastal areas, to meet the relative insufficiency of the labour force there. According to the World Bank's estimates, in the late 1990s the average annual growth rate of demand for labour in rural areas will be about 0.8 per cent. All new employment opportunities will be created by non-agricultural sectors in rural areas. In contrast, the average annual growth rate of demand for labour in urban areas will be about 2.8 per cent, which is much higher than in rural areas (World Bank, 1992).

The differences in labour demand and supply between rural and urban areas indicate the existence of abundant surplus labour in most rural areas on the one hand, and a shortage of labour supply in some urban areas on the other. It is thus important that regional imbalances in the demand and supply of labour are adjusted by moving rural labour to urban areas. Rural workers moving to urban areas gain the advantage of increasing incomes while urban areas benefit from an improved labour supply. In 1993 the annual average number of rural labourers moving to cities was about 50 million, of whom 20 million were moving across provinces (*Xinming Wanbao*, 1994).

The differences between urban and rural areas have been seen to be widening in several aspects. In terms of income level, urban income per capita was 2.2 times rural income in 1964 (Volume 7 of Study, Research, Reference, 1993). It had increased to 4 times as great in 1991. From a social security perspective, urban citizens have entitlements to pensions and medical treatment expenses as long as they are employed. By contrast, rural people have no such benefits and also may have to provide for a range of family social expenses. In 1991 urban social insurance expenditure was 30 times as great as rural. In relation to housing, urban citizens enjoy government housing subsidies and have better housing conditions, such as clean water, gas and toilet facilities. In rural areas, farmers have to build their houses themselves, causing some to get into debt. Some houses are built from rough materials such as mud and grass. About 90 per cent of rural households use wood and grass as the major fuel. In terms of transportation and postal communications, at least 8 per cent of rural areas have no highway, and 30 per cent of the townships and towns have no post offices, while 2.8 per cent of the townships and 57 per cent of the villages have no telephone facilities. In education, the enrolment rate of school-aged children is about 99 per cent in cities and towns, about 80 per cent in rural areas and about 50 per cent in remote mountain and pasture areas. The rate of education after primary school graduation is almost 100 per cent in urban areas, and 59 per cent in rural areas. The rate for pupils going up a level after junior middle school graduation is 69 per cent in urban areas, but only 10 per cent in rural areas. About 24 per cent of the counties that have not universalized nine-year compulsory education. From the perspective of health care, most urban citizens benefit from the public-funded health care system, and the average medical expenditure of an employer reaches 180 yuan. In rural areas in 1991, however, about one third of townships had not established a hospital, and 100 thousand villages did not even have clinics. In terms of cultural life, many counties have no theatres, cinemas or libraries. About half of the townships have no place offering cultural entertainment and 20 per cent of rural areas are not covered by radio broadcasts.

As we know, the social inequalities in China are not only a result of reform. Their pattern and scope have long historic roots, changing over time. With the rapid development of the market economy the income gap between urban and rural people has become larger. This is one of the major driving forces encouraging rural labour to move to urban areas. Statistics show that urban per capita income was 2.24 times greater than rural in 1980. The gap was reduced to 1.9 times in 1985, but increased to 2.5 times in 1993. This zigzag change was due, to a great extent, to the restraints imposed on the free movement of production elements (the labour force in particular) by the government. The current difference in income between rural and urban dwellers has exceeded the levels of developing countries and other areas in Asia.

Generally, the ability to absorb the floating population is much greater in economically developed regions than in less developed areas. According to statistics, there are eight provinces and municipalities whose national income per capita is over 1,500 yuan a year, namely Shanghai, Beijing, Tianjin, Liaoning, Guangdong, Zhejiang, Jiangsu, and Heilongjiang. The population of these eight provinces and municipalities is 278 million. They absorb about 13.68 million floating population, and the average floating population per thousand of the local resident population is 49. For those provinces and autonomous regions whose national income per capita is between 1,000 and 1,500 yuan a year, such as Jilin, Shandong, Fujian, Shanxi, Ningxia, and so forth, the average floating population per thousand local resident population is 25.8. For those regions whose national income is less than 1,000 yuan a year, the average floating population per thousand local resident population is only 17 (Liu Futan, 1992).

Change of land resource conditions and migration

The major cause of the large amount of floating population in China is the existence of a considerable rural surplus of labour. This surplus labour in rural areas is a result of the huge size of the total population and the high proportion of agricultural population. This situation is summed up in the expression describing China as a country of 'many people with little land'.

Historically, the total population has increased constantly and arable land per capita decreased dramatically. According to statistics, the population of China increased from 0.646 billion in 1957 to 1.185 billion in 1993. During the same period, the rural population increased from 547 million to 852 million, and the rural labour force increased from 193 million to 340 million. The total area of arable land decreased from 111.3 million hectares to 95.3 million, a reduction of 14.4 per cent. This resulted in a significant decrease of arable land per capita from 0.17 hectares in 1957 to 0.08 in 1993. At the same time, the arable land per capita of the rural labour force decreased from 0.58 hectares to 0.28.

The national economic structure of China has experienced a series of adjustments. First, the relative sizes of the agricultural and non-agricultural population have changed rapidly. The proportion of agricultural population declined from more than 85 per cent in the early 1950s to less than 70 per cent in 1993. Second, the share of industrial output value increased from 43 per cent to 83 per cent, and that of agricultural output dropped from 57 per cent to 17 per cent. Third, the proportion of urban population increased from 11 per cent to near 30 per cent.

Considering the above situation, the total rural surplus labour force was estimated to be at least 176 million at the end of 1993. An issue of great concern is that the annual increase in the rural working-age population reached 5 to 10 million recently. Given this situation of population development, many working-age people in rural areas will face uncertain employment prospects.

Population policy and the increase in floating population

There are a number of factors that account for the dramatic increase in floating population including socio-economic conditions, government policy and demographic factors. The implementation of a family-planning policy has played an important role in stimulating a decline in fertility levels in China since the early 1970s. The average number of children per woman decreased from 5–6 in the 1950–60s to 2–3 in the later 1970s. However, the decline of the total fertility rate is unevenly distributed between urban and rural areas, and from region to region. Nowadays most urban families have only one child. However, two- or more-than-two-children families are still common in the countryside, particularly in poor remote rural areas. This is strongly influenced by the traditional views about birth, reflected in sayings such as 'the bigger the number of children, the bigger your fortune', 'raise children for old-age support' and 'propagate to continue the family line'. Another influence is the rural household responsibility system which, being based upon the family as the basic unit, stimulates rural families to have more children: the more children, the more family labourers, and the more wealth. According to statistics, the natural growth rate of population in urban areas was 9.38 per thousand in 1993. It was 12.17 per thousand in rural areas, which marked a significant re-increase since the implementation of the household responsibility system (*China Statistical Yearbook*, 1994). From a long-term perspective, the birth peaks will turn out to be the peaks of labour supply. Recently, this has been witnessed by a faster increase of the labour force in rural areas than in urban areas.

The root causes of the floating population are gaps in demographic transition, differences of social and economic level and income disparity. The lag in demographic transition and the surplus of labour in rural areas push young people to leave. On the other hand, the rapid decline of urban fertility and ageing population means that urban areas face labour shortages and provide opportunities for rural labour. At the micro level, income disparity is the most significant factor and becomes the major attractant drawing in rural migrants. Furthermore, the relaxation of policy makes the possibility of floating population flow become the reality.

The causes of rural migration are not only on the supply side (surplus labour), but also due to the demands of urban economic development. Surplus rural labour has become an important real as well as potential motor in the urban economy. Rural migrants have an important position in the labour market, including in the informal sector. The migration of rural labourers has both changed and charged the economic dynamics of China's large urban centres, such as Shanghai. It is clear that migrants from rural areas are a major labour supply for the basic production and service industries which are part of the growing urban economy. In Shanghai , as well as other urban areas, the extraordinarily low fertility of the past two decades and the booming economy will no doubt entail a continued shortage of labour and a continued need for rural migrants. Urban labour demand opens up opportunities for rural Chinese for new economic and social mobility. Rural migrant labourers make up a low-cost and low-income group in urban areas, thus benefiting competitive production. Urban areas also gain from an effective influx of low-cost human capital.

However, rural to urban migration, while occurring on a large scale in China, has not so far resulted in a clear direction of economic or social integration between urban and rural areas, and between urban and rural Chinese. Rural migrants, such as those in Shanghai, come to the bright city lights to improve their lives, but in the minds of most there is no intention to move to the cities. They migrate, in other words, to work, but not to live. Rural migrants working in urban Shanghai, have maintained their 'rural' identity. Their occupational structure is distinctively different from that of local residents and their living environment is also distinctively different from that of urban Shanghainese. Only 3 per cent are in professional, cadre or clerical positions and the overwhelming majority come to Shanghai to work as manual labourers, and service providers. Such a tiny percentage is in sharp contrast with the over 25 per cent of urban employees working in such occupations. While rural migrants can come to cities to make a living, the extensive social welfare system is still available only to urban residents. Most rural migrants, therefore, have not resettled in Shanghai. They are 'floating on the surface' of Chinese urban society, not living in it. Their housing arrangements are mostly temporary, family life is at best transitional, and access to basic benefits such as medical, labour insurance, and old age pension is almost non-existent.

There is a need for further research on the possibility of establishing an integrated labour market in urban areas and tackling the discrimination faced by rural labourers in employment. The design and implementation of pilot projects providing services for rural migrants, especially those aimed at improving their living conditions, are also a potential area for future work. Thus in the future floating rural migrants may settle in urban areas, on a more permanent, secure basis.

References

Ai, Xiao (1989), Current floating population reaches 50 million in China. *People's Daily*, 4 March.

Editorial Committee (1988), *Dangdai Zhongguo Renkou* (Contemporary Population in China), Beijing: China Social Sciences Press, p. 255.

Fazhi Ribao (Legality Daily), 7 June 1994.

Gong, Shengzhu (1991), Comparison of the two patterns of China's population migration, *Zhongguo Renkou Kexue* (China Population Sciences), 1991 (4).

Ji, Dangsheng et al. (1995), Present conditions of the floating population in Beijing and countermeasures research, *Zhongguo Renkou Kexue* (China Population Sciences), 1995 (4).

Li, Chubai (1991), *Influences of the Floating Population on Large Cities and Countermeasures*. Beijing: Economy Daily Press (In Chinese).

Liu, Futan (1992), *Nongcun Gaige De Xinzhanlue* (New Strategy of Rural Reform), China Finance and Economy Publishing Service, p. 15.

Liu, Xiuhua (1996), *Analysis of Floating Population in Beijing*. Beijing Statistics Bureau (in Chinese).

National Statistics Bureau (1991), *China Statistical Yearbook, 1990*. China Statistical Press.

National Statistics Bureau (1995), *China Statistical Yearbook, 1994*. China Statistical Press.

Research Group of Rural Annual Analysis (1993), *Annual Report of Rural Economic Development in China and Forecast Development Trends*. Beijing: China Social Sciences Press (in Chinese).

Shanghai Economic Yearbook 1994, Shanghai Economic Yearbook Press.

Shanghai Statistics Bureau (1994), *Shanghai Statistical Yearbook, 1994*. China Statistical Press.

Shanghai Statistics Bureau (1995), *Shanghai Statistical Yearbook, 1995*. China Statistical Press.

Wang, Wuding, Le, Huizhong and Zuo, Xuejin (eds) (1996), *Shanghai's Floating Population in 1990s*. Shanghai: The East China Normal University Press.

The World Bank (1992), *China: Strategies for Reducing Poverty in the 1990s*, Washington DC: The World Bank.

Xinming Wanbao (Xinming Evening News) 20 February 1994.

Zhang, Tiejun (1996), *Analysis of Floating Population in Beijing*, Beijing Statistics Bureau, January (in Chinese).

Zhao, Changbao (1994), Economic development and mobility of rural labour force investigation and analysis of the present out-moving rural labour force. In *Investigation and Research on the Wave of Farmer Workers*, Rural Economy Research Centre of the Ministry of Agriculture.

15   International Migration Patterns

Ye Wenzhen

International migration in China has several important features. First, China has a very long history of international migration, which is said to have begun more than two thousand years ago during the Qin Dynasty. Second, the numbers of people involved in international migration have been enormous. It has been estimated that there are about 37 million overseas Chinese in the world at present and they are widely distributed in 136 countries and areas in the five continents (Poston et al., 1995). Third, the migration rate is extremely low, approximately only one per cent in recent times (Zhang, 1992), because of the huge size of the Chinese population. Finally, China has been a net exporter of people over history. Based on the stage theory of China's international migration (Zhu, 1987a), this chapter attempts to offer a general survey of international migration in China, with a special attention given to its patterns and major causes.

Beginning Stage: Who was the First International Emigrant?

The first stage of international migration in China lasted from the Qin to the Tang Dynasty. According to many scholars, the earliest international emigrant in Chinese history was Xu Fu, from Shangdong Province. In 219 BC, Xu Fu was asked by the Qin Emperor to go by ship, with thousands of children, to Japan to search for medicinal crops. However, Xu Fu decided to stay in Japan instead of going back to China (Zheng et al., 1985; Zhu, 1987b; Wang and Du, 1992). Thereafter, many diplomats were sent by the emperors from the Han to the Sui Dynasties to western countries such as Cambodia and Thailand, merchant ships went to Myanmar, India, and Sri Lanka through South China Sea and Chinese monks and priests landed in India for religious purposes. At this stage, the numbers of people involved in international movements were small and increased very slowly. Since most of these movements were motivated by diplomacy and trade, very few migrants stayed abroad permanently.

Spontaneous Stage: A More Business-Related Migration Behaviour

The period from the Tang and Song Dynasties to the opening of sea areas in the middle and late Ming Dynasty is the second stage of the history of international migration in China. The numbers of people going abroad increased during this period. More and more young students, in addition to the merchants, diplomats, and monks, began to travel internationally in order to study for advanced degrees in foreign countries. However, merchants were the dominant component of the population flow. Some scholars, therefore, refer to this stage of international migration in China as an emigration of Chinese merchants to foreign countries (Wang, 1991). The mode of the movement also changed – from an appointed or visiting behaviour to a spontaneous or personally motivated one and to a group shift. As a direct result of the increasing scale of international migration, the number of overseas Chinese experienced sustained growth and finally reached several tens of thousands by the end of the stage (see table 15.1). Most of them concentrated in the southeast Asian countries, such as Indonesia, Singapore, Malaysia, Vietnam, Thailand, and the Philippines, which were the traffic and trade centres of the time.

The major reasons for the significant growth in the numbers of international migrants at this stage, according to Li's analysis, were the expansion of international trade and the developments in maritime transportation in China at that time (Li, 1986). Beginning in the Tang Dynasty, China had made continuous progress in its navigation, its invention and use of the compass being one of the best-known technical advances made in the ancient world. At the same time, shipbuilding in China was also flourishing. While still using the Silk Road of the Han Dynasty, China had opened up many new international sealanes connecting it with more and more foreign countries. With the aid of advanced maritime transportation techniques and instruments, a significant increase in international trade occurred in China: Guangzhou, Quanzhou, and Mingzhou becoming the most thriving open ports in the Nansong dynasty; the number of major ports increasing to seven when China entered the Yuan Dynasty; and Zhenghe's great fleet dominating the oceans during the Ming era. It was the interaction between the development in maritime transportation and the growth in international trade that brought about the first large wave of international emigration in Chinese history.

Push and Pull Stage: A Massive Labourer Migration

After the Ming Dynasty, China entered the third stage of its history of international migration, which was to last until the People's Republic of China was founded in 1949. As millions of men went abroad to work overseas in this period, this stage is often historically recognized as a time when a massive emigration of Chinese labourers took place.

Table 15.1 The growth of the overseas Chinese
population (in 1,000s)

Year	Population	Year	Population
up to Ming		1962	16,359.2
Dynasty	<100	1963	16,897.1
Qing Dynasty		1964	17,428.6
before 1840	100–1,000	1965	17,563.4
1879	3,000	1966	17,736.3
1899	4,000	1967	18,118.4
1905	7,600	1968	18,298.0
1921	8,600	1969	18,800.3
1931	8,480	1970	19,293.8
1940	8,500	1971	19,833.8
1948	8,721.2	1972	20,234.5
1949	10,719.0	1973	21,063.4
1950	11,093.9	1974	21,466.5
1951	12,127.0	1975	22,025.5
1952	12,536.2	1976	22,588.1
1953	13,330.2	1977	23,202.6
1954	13,472.3	1978	24,037.3
1955	14,126.7	1979	24,472.9
1956	14,207.7	1980	24,653.5
1957	14,405.0	1981	25,583.7
1958	14,471.7	1982	26,092.0
1959	14,581.1	1983	26,195.7
1960	15,385.2	1990	36,765.8
1961	15,709.7		

Sources: Zheng et al. (1985), *Overseas Chinese*, p. 202,
Beijing: People's Press; Poston et al. (1995), The distribution of overseas Chinese in the contemporary world,
Population Sciences of China, 95 (1), 9–15; Poston et al.
(1990), The distribution of overseas Chinese in the contemporary world, *International Migration Review*, 24
(3), 480–506.

Three ways to go abroad

There were three ways for these labourers to leave for foreign countries.
The first was privately funded migration, i.e. they managed to pay for their
own passages by borrowing money from friends and relatives, or by selling
or mortgaging farms and other personal and family possessions.

The second way was the credit-ticket system, under which funds for the
passage were advanced by a sponsor, such as a merchant or a company, to an
emigrant who agreed to pay them back in the ensuing months after arrival.

The third way was the contract labour system. This required the emigrants to sign contracts agreeing to work in the destination country for a specified time in return for passage. Under the contract system labourers were frequently kidnapped or tricked and coerced into going abroad. Many died due to the extremely difficult living conditions on board ship or the failure of organized mutinies before they reached their foreign destinations. For example, in 1855, nearly 300 of the 450 'coolies' aboard the American ship *Waverly* travelling from Swatou to Callao, Peru, died of suffocation (Mark and Chih, 1982). It was estimated that the travelling mortality rate of 'coolies' from China to Cuba, Peru, Panama and America ranged from 22 to 64 per cent during the period 1847–73 (Zhu, 1987a). Upon their arrival they were often forced to work under slave-like conditions. The mortality rate of Chinese labourers within the contract period was as high as 75 per cent and the average working-life expectation was only five years (Ye and Lin, 1994). Because of numerous accounts of inhumane treatment and a very high rate of death, the contracting system became known as 'pig-selling' by the Chinese and 'coolie trading' by foreigners (Thernstrom, 1980).

Emigration patterns and features

According to some Chinese scholars' estimates (Peng, 1980; Wang and Du, 1992), about 10 million Chinese labourers went abroad during the period of about two hundred years to the middle of the twentieth century. But the more often cited numbers of Chinese labourers leaving the country and their distributions in the different foreign destinations in the years from 1800 to 1925 are from Chen's study (Chen, 1963) and shown in table 15.2. Over a period of 125 years, about three million Chinese labourers were sent abroad. More than two thirds of them went to southeast Asian countries, while one sixth chose America as their destination. It is very clear from the table that the largest wave of Chinese labour emigration occurred in the

Table 15.2 The estimated numbers of Chinese labourers going abroad (1,000s)

Years	Regions and countries								
	Total	Asia	Australia	USA	Canada	Cuba	Peru	Europe	Other
1801–1850	320	215	10	18		17	10		50
1851–1875	1,280	675	55	160	30	135	110		115
1876–1900	750	700	8	12	4				26
1901–1925	650	425						150	75
1801–1925	3,000	2,015	73	190	34	152	120	150	266

Sources: Chen, Zhexiang (1960) The contract system of Chinese labourers in the 19th century. *History Studies*, 63 (1) (in Chinese).

third quarter of the nineteenth century, with 55,000 labourers leaving the country every year.

Most of these Chinese labour emigrants were from Fujian and Guangdong with limited numbers also from Shanghai and Zhejiang Provinces, all located in the southeastern part of China.

Compared with the first two stages, several interesting differences in the patterns can be revealed based on the historical records from the treaty ports, such as Xiamen, Shangtou, Guangzhou, Fuzhou, Ningbo, Marco and Hong Kong:

1. All of them were the most able-bodied males in their home villages or towns whose ages ranged from the early twenties to the late thirties. They first left their home town to find employment in the cities and ports. Once there, they enlisted for work as unskilled labourers and finally became the major target population to be sent abroad as 'coolies'.
2. A high percentage of them came from a lower social and economic status. They were poorly educated and most of them worked, prior going abroad, as peasants, fishermen, small vendors and unskilled workers.
3. The numbers of female emigrants apparently increased in this period. For example, only 16 Chinese women were sent to San Francisco before 1854 while about 45,000 Chinese male emigrants were there at that time. But by the 1860s, the percentage of female emigrants in the treaty port, Hong Kong, had increased to 5.7 per cent (Wu, 1988). And the total number of Chinese women leaving Hong Kong for foreign countries exceeded 100,000 during the last 40 years of the nineteenth century (Wu, 1988).
4. Most of them stayed on at their destination to work and live as overseas Chinese. By the end of the 1930s, the size of the overseas Chinese population was almost 9 million (see table 15.1) due to the large numbers of Chinese labourers going and staying abroad.

Causal factors of labour emigration

The causes of Chinese labour emigration can be explained by Lee's 'push-and-pull' theory (Lee, 1966). The push factors which are associated with the place of origin included drought and floods, mass famine, official corruption, peasant revolts and the western nations' invasion during the nineteenth century, the combined effects of which led the Qin Dynasty to fall into decay and put China in crisis. In 1846 and 1848, the population was reduced to desperation by massive starvation and poverty caused partly by drought and floods which ruined staple crops. After the Opium War continuous economic and political pressure exerted by the western invaders led to a rapid breakdown in China's economy and traditional society and brought still greater hardships to the masses. In mid-century the Taiping Rebellion shook the empire further to its foundations (Thernstrom, 1980). Struggling for survival during this troubled period, many people, especially in the southeast, were forced to leave. Political conflicts, economic crises and social problems in China worked together to push people to emigrate (Wu, 1988).

On the other hand, changes elsewhere in the world in the mid-nineteenth century also affected the patterns of Chinese emigration. Those so called 'pull' factors were the manpower needs created by the discovery of gold in the United States, Canada, Australia and New Zealand; the era of coast-to-coast railroad building in North America; the beginning of the construction of the Panama Canal; and the further development of economic plants by western European nations in their colonial areas. The impoverished Chinese peasants were believed to be the cheapest and most industrious labour force available to fulfil these needs.

In addition, a number of factors played an intermediary role facilitating this international migration. The historical accumulation of migration experience, the information about the destination countries diffused by former emigrants and the examples of people getting rich in foreign nations all encouraged emigration.

New Stage: A More Policy-Related International Movement

Since 1949 China's international migration has moved to a new stage that has several important patterns and structural features. During this stage, the governmental policies of both sending and receiving countries toward immigration have become a dominant factor affecting the patterns of Chinese international movement.

Two-way international flows

The first pattern is reflected by a change from a one-way to a two-way movement, indicating an increasing size of population inflow. The economic recovery from the end of the civil war and the further stabilization of social life during the early 1950s attracted a great number of overseas Chinese to come back for various purposes, such as to participate in the construction of a socialist state, to study for degrees, to retire or to be reunited with their families. During this, the first wave of international immigration in China's history, the estimated yearly number of returning overseas Chinese and their children was more than 10,000 (Zhu, 1987a).

Later another two massive inflows of overseas Chinese were recorded in 1960 and the late 1970s, respectively, partly due to the changes in foreign policies associated with China in the destination countries. In 1960 the Chinese exclusion conducted by Indonesia's government not only ruined thousands of overseas Chinese families, but also forced many of them to leave for their motherland. A similar outflow of Chinese occurred when the diplomatic relationship between Vietnam and China broke down in the late 1970s. According to an incomplete estimation, the number of refugees resulting directly from the exclusion of Chinese in Vietnam was as many as two million, most of whom were overseas Chinese (Zhu, 1987a).

Since the early 1980s when the reform and 'open-door' policy was adopted in China, the abundant career-developing and money-making opportunities created in coastal areas and other open cities have attracted a great number of overseas Chinese and foreign business interests to mainland China. Many of them have even applied for and finally received a green card with which they can stay in China as permanent residents. Furthermore, in 1988, there were about 70,000 foreign experts and scholars working for universities, research institutions and industrial and mining enterprises in China. In addition, about 2 million foreign tourists visited China in the same year (Wang and Du, 1992). A more interesting fact is that many foreign labourers cross the border in Heilongjiang, Yunnan and Guangxi Provinces to find employment opportunities and enjoy better living conditions in China even without legal documentation. It seems that China, once characterized by a number of 'push' factors provoking people to leave, is now a place pulling in more and more emigrants.

An increasing outflow for a variety of purposes

The second pattern of current international migration in China is that the population outflow has also continuously increased except for a policy-related cessation during the 1966–67 Cultural Revolution period. During the first 15 years (1949–65) of socialist China, international emigration was still an important demographic phenomenon, especially in Fujian and Guangdong Provinces, although the volume of outflow population was smaller than that of the inflow. Most of the outflow was related to foreign marriages, family reunions, and the inheritance of overseas property, which clearly means that emigration in this period was directly associated with the earlier international movement. The age and sex structures were also different. More and more children, women, and elderly people were involved in the demographic shift.

After 1976 China began to see a massive population outflow. According to administrative records, the number of Chinese citizens going abroad for personal reasons increased steadily from about 40,000 in 1986 to 53,995 in 1987, 128,354 in 1988, and 132,727 in 1989, with an average annual growth rate of 49.2 per cent from 1986 to 1989 (Wang and Du, 1992). In 1993, about 33,000 Chinese citizens left or were waiting to leave for other countries by establishing a marital relationship with foreigners (Ye and Ling, 1996). Recently, more than 10,000 scholars from universities and research institutions, administrators at all levels, and other professionals from industrial and mining enterprises have been sent annually to foreign nations for advanced training and studies. And about 35,000 labourers of all kinds have worked abroad temporarily.

The current outflow has shown three trends: emigrant-sending areas have extended from the traditional provinces of Fujian and Guangdong to other coastal and even inland cities and provinces; more and more Chinese emigrants are moving to North America, especially the United States, and to Australia, simply because of the greater possibility of remaining there longer or permanently with legal status; and there has been higher selectivity in migration with the better-educated, professional and relatively younger cohorts as the target population.

There are several factors contributing to this new wave of emigration in Chinese history. First, during the Cultural Revolution, China insisted on a closed political and economic policy, which gave very limited chances for its citizens to go abroad. People found that it was not only extremely difficult to get their exit application approved, but also dangerous, because at the time any overseas connection could result in political trouble for the individual concerned. With this political orientation, almost all emigration, even when it had a reasonable basis, was stopped and the number of individuals leaving for foreign countries dropped to nearly zero. The new wave of emigration, therefore, is partly due to international movements 'catching up' as a result of a significant change in foreign policy, namely from a 'closed' to an 'open' policy.

The reform and open-door policies starting in the early 1980s not only freed people's minds of political apprehension about going abroad but also provided them with abundant opportunities to be in contact with foreigners in economic, cultural, social, academic and other contexts. When a relationship network has been developed through frequent contact with outside world, a demographic shift becomes an inevitable social phenomenon. So it is reasonable to believe that the policy factor is the most important force in the resumption of Chinese emigration to foreign nations.

Secondly, the diversification of the value system in Chinese society during the 1980s and 1990s is also considered to be another important contributor to the outflow wave. For many Chinese, worship of and blind faith in everything associated with foreign nations, especially the West, drive them to take any possible measures to leave the country. Marrying a foreigner is a method now employed by more and more young Chinese people to realize their dreams of going abroad (Ye and Ling, 1996). Moreover, the recently emerging 'money culture' and the display by some visiting overseas Chinese of the wealth they have accumulated in a few years overseas have further intensified 'West worship' among the masses. Starting approximately in 1978, many people who left earlier began to return to Fujian, Guangdong, and other places of origin for family visits. Once they arrived in their ancestral home, they stayed in fancy hotels, gave expensive gifts to their relatives, began buying or building huge and luxurious houses, and even donated large sums of money to their communities to construct classroom buildings for

children and provide facilities for the elderly. The fact that returning overseas Chinese make their wealth conspicuous has made a deep impression on
their fellow countrymen, further encouraging them to try to emigrate themselves. Furthermore, people believe that they too could become rich in the
outside world because the visiting overseas Chinese came from similar socio-
economic backgrounds to their own before they went abroad.

The immigration policies adopted by some destination countries are the
third factor facilitating emigration. The priority given to family members
and professionals by the immigration laws in many western countries
has been responsible both for the high selectivity in emigration and
the enlarged scale of population outflow in China. At the same time, the
political considerations that gave rise to the immigration laws strongly
encourage undocumented emigration.

Secondary movements of overseas Chinese

The third international migration pattern is associated with the secondary
movement of overseas Chinese who change their original country of destination for another foreign nation (Wang, 1991). The movement of overseas
Chinese residing in southeast Asian countries to western Europe and of
Chinese residents in Hong Kong and Macao to North America during
recent decades reflects the fact that serial migrations will be made by overseas Chinese if necessary.

Structural features of current outflow

Compared with international movements during the previous three stages,
the current outflow population has several different structural features.
Using recent Chinese immigrants to the United States as an example, it
was first found that the number of female Chinese immigrants was greater
than males. Historically, Chinese men migrants outnumbered women: 14 to
1 in 1910, 7 to 1 in 1920, and 2 to 1 in 1950 (Hing, 1993). By 1988, however,
the margin had been reversed; there were 47.2 men to 52.8 women among
every 100 Chinese immigrants to the United States, indicating that there
are more women now participating in the international flow.

Secondly, the age distribution of the outflow population is getting more
and more balanced. For instance, also in 1988, the percentages of Chinese
immigrants in the 0–19, 20–39, 40–59, and 60+ age groups were 18.6, 36.5,
27.6, and 17.3, respectively, reflecting that young children and elderly people
are also becoming an important part of the international outflow.

Third, as discussed above, traditional Chinese emigrants were poorly
educated, non-professional and from families with a lower socio-economic
status. But a significant change has occurred in recent times in the personal
and family backgrounds of Chinese emigrants. According to records from
the INS of the United States in the fiscal year of 1988, about 32 per cent of
Chinese immigrants of working age (from 20 to 59) had professional exper-

tise and technical, executive, administrative, managerial or administrative support occupations when they arrived. 15.4 per cent had work experience in sales, service, precision production, crafts, and repairs. These occupational groups generally have a better education and professional training and come from urban areas. Therefore, it is reasonable to say that the outflow Chinese population in recent years is relatively highly qualified.

Demographic consequences of international migration

According to the disruption hypothesis (Kahn, 1991; Stephen and Bean, 1992), the assimilation theory (Ford, 1990), and the minority group status perspectives (Espenshade and Ye, 1994), Chinese emigrants have a low level of fertility partly due to financial hardship, psychological stress, and the separation of spouses caused by a move, they show a relatively high level of assimilation to the dominant society and culture, and they make sacrificial efforts in their pursuit of social and economic equality as a minority. For example, 'as measured in the 1980 census, the average number of children ever born for Chinese–American women aged 15 and older was 1.65, which was not only less than the number for other Asian–American women (1.75), but also less than the average of 1.82 recorded by white females' (Espenshade and Ye, 1994). Therefore, the total number of the overseas Chinese population and its distribution over the world is more likely to be determined by the patterns of immigration. From Table 15.1 we can see that the number of overseas Chinese significantly increased from 12,536,200 in 1952 to 36,765,800 in 1990. The average annual growth rate over those 38 years was 2.9 per cent. The highest rate was during the years 1975–90 (3.48 per cent) and the lowest in the period 1963–75 (2.15 per cent), which correlates strongly with the changes in the volumes of population outflow in China since the early 1950s.

Table 15.3 shows the changes in the distribution of overseas Chinese population in the world during the period from 1952 to 1990. In 1952 about 98 per cent of the overseas Chinese population were concentrated in Asia

Table 15.3 Overseas Chinese population in the five continents (1,000s), 1952–1990

	1952		1960		1970		1980		1990	
	No.	%	No.	%	No.	%	No.	%	No.	%
Total	12,536.2	100	15,385.2	100	19,293.8	100	26,972.4	100	36,765.8	100
Asia	12,228.5	97.5	14,880.1	96.7	18,342.6	95.1	24,764.0	91.8	32,287.8	87.8
America	203.9	1.6	406.6	2.6	711.2	3.7	1,333.0	4.9	3,226.6	8.8
Europe	11.5	0.1	15.8	0.1	112.1	0.6	622.0	2.3	769.5	2.1
Oceania	60.9	0.5	42.1	0.3	68.5	0.4	176.4	0.7	373.9	1.0
Africa	31.3	0.2	40.6	0.3	59.3	0.3	76.9	0.3	108.0	0.3

Sources: Poston et al. (1995), The distribution of overseas Chinese in the contemporary world. *Population Sciences of China*, 95 (1), 9–15.

while only two percent lived in the other four continents. Since then, the percentage of overseas Chinese in Asia has continuously declined from 97.5 per cent in 1952 to 96.7 in 1960, 95.1 in 1970, 91.8 around 1980, and 87.8 around 1990, which was probably due to fewer and fewer people from China moving to other Asian countries and more and more Chinese emigrants in Asia going to America or Oceania or returning to China during this period of time. The percentage of Chinese population in America and Oceania, however, has kept rising from 1.6 per cent in 1952 to 8.8 around 1990 for the former, and from 0.3 in 1960 to 1.0 per cent around 1990 for the latter. At the present time, among every 100 overseas Chinese, 88 are living in Asia, 9 in America, 2 in Europe, and only 1 in Oceania. But it is reasonable to believe that if the changes in the distribution of overseas Chinese population during the last four decades continue, more and more Chinese will be found in America and Oceania in future.

Table 15.4 shows the distribution patterns of the overseas Chinese population by individual countries that had a Chinese population of ten thousand or above in the 1980s. In 1980 Indonesia, Thailand, Hong Kong, Malaysia and Singapore accounted for about 80 per cent of the total overseas population in the world, indicating that four out of every five overseas Chinese were living in these five Asian nations. Among them, Indonesia was the country with the largest number of overseas Chinese (6.15 million, accounting for 23 per cent of the total). By the end of the 1980s, however, the percentage of Chinese population in these five countries had fallen to 72.3. Indonesia was still the number one country in terms of the size of its Chinese population (7.32 million), followed by Thailand (6 million), Hong Kong (5.69 million), Malaysia (5.47 million), Singapore (2.11 million), Vietnam (2 million), respectively. The United States was the seventh in order (1.65 million). While Laos experienced the highest average annual growth rate of Chinese population in the 1980s (39.6 per cent), the Philippines had the lowest (–2.6 per cent).

Summary

In this chapter, the patterns and features of the Chinese international migration experience have been carefully outlined. By 1990 the total number of overseas Chinese in the world had reached as high as 36.77 million, which was believed to be mainly due to a continuous outflow of Chinese population over the last several hundred years. The results from the historical analysis show us at least four important shifts in the patterns of international migration in China:

1.   The target outflow population was no longer concentrated in the traditional areas of Fujian, Guangdong, and other coastal provinces in the southeastern part of China. Urban residents are more likely to go abroad.

Table 15.4 The distribution of the overseas Chinese population in selective countries (1,000s), 1980–1991

	Around 1980		Around 1990		Average annual growth (%)
	Number	%	Number	%	
Total	26,972.4	100.0	36,765.8	100.0	2.7
Asia	24,746.0	91.8	32,278.8	87.8	2.4
Indonesia	6,150.0	22.8	7,315.0	19.9	1.9
Thailand	4,800.0	17.8	6,000.0	16.3	2.0
Hong Kong	4,885.6	18.1	5,686.1	15.5	1.5
Malaysia	3,630.5	13.5	5,471.7	14.9	3.7
Singapore	1,856.2	6.9	2,112.7	5.7	1.2
Vietnam	700.0	2.6	2,000.0	5.4	13.1
Myanmar	700.0	2.6	1,500.0	4.1	8.5
Philippines	1,036.0	3.8	820.0	2.2	−2.6
Macau	271.1	1.0	423.7	1.2	4.5
Cambodia	360.0	1.3	300.0	0.8	−0.9
Laos	10.0	0.0	160.0	0.4	39.6
Japan	54.6	0.2	150.3	0.4	12.7
India	110.0	0.4	130.0	0.4	1.9
America	1,333.0	4.9	3,226.6	8.8	7.9
USA	806.0	3.0	1,645.6	4.5	7.1
Canada	289.2	1.1	680.0	1.8	8.5
Peru	52.0	0.2	500.0	1.4	25.1
Brazil	11.0	0.0	100.0	0.3	19.9
Panama	33.0	0.1	100.0	0.3	15.8
Europe	622.0	2.3	769.5	2.1	3.2
Soviet Union	222.0	0.8	274.0	0.7	6.7
France	210.0	0.8	200.0	0.5	−0.6
UK	91.0	0.3	125.0	0.3	6.3
Oceania	176.4	0.7	373.9	1.0	8.5
Australia	122.7	0.5	300.0	0.8	11.2

Sources: Poston et al. (1995), The distribution of overseas Chinese in the contemporary world, *Population Sciences of China*, 95 (1), 9–15.

2. The age and sex distribution of outflow population was no longer skewed towards young and male Chinese as was the case with the primary emigrants. Children, women and old people have also become a major component of international movements in contemporary China. The qualification level of the modern emigrants has significantly improved.

3. Southeast Asian countries have progressively lost their attractiveness as destinations for Chinese emigrants. More and more Chinese

go to developed countries such as North America, Australia, and Europe.

4. China will receive an increasing inflow population during next several decades, partly due to its faster economic growth and greater demand for an educated foreign labour force.

Furthermore, the findings strongly indicate that in addition to the economic, social and environmental factors, changes in the government immigration policies adopted by both sending and receiving countries have made a significant contribution to the evolution of China's international population flows. For example, without mainland China's open-door policy that allows its citizens to go abroad for different purposes, such as to visit families and friends, find jobs, settle and study or travel, there would not have been such a dramatic increase in the size of the outflow population during the 1980s.

At the same time, the immigration policies of the host countries have also had an impact on the numbers of Chinese entering their countries. During the 1980s, for instance, the growth rate of overseas Chinese in the

Table 15.5 The effects of US immigration policies upon immigration flow from China and the Chinese–American population: 1860–1890

Decade ending	Population	Immigration in prior decade*	Law in effect in prior decade
1860	34,933	41,397	open
1870	64,199	64,301	open until Burlingame Treaty in 1868
1880	105,465	123,201	Burlingame Treaty
1890	107,488	61,711	Burlingame Treaty until 1882, then Chinese Exclusion Act
1900	118,746	14,799	Chinese Exclusion Act
1910	94,414	20,605	Chinese Exclusion Act
1920	85,202	21,278	Chinese Exclusion Act
1930	102,159	29,907	Chinese Exclusion Act and 1924 Act
1940	106,334	4,928	Chinese Exclusion Act and 1924 Act
1950	150,005	16,709	Chinese Exclusion Act, then Chinese Exclusion Act repealed in 1943
1960	237,292	25,201	Chinese Exclusion Act repealed, then Asia-Pacific triangle in 1952
1970	436,062	109,771	Asia-Pacific triangle, then 1965 amendments
1980	812,178	237,793	1965 amendments
1990	1,645,472	446,000	1965 amendments and separate Taiwan quota

Sources: Hing, Bill Ong (1993) *Making and Remaking Asian America through Immigration Policy:1850–1990*, p. 48, Stanford: Stanford University Press.

Americas and Oceania was two or three times higher than that in other continents (8 and 8.5 per cent for the Americas and Oceania, respectively; 2.3, 2.4 and 3.2 per cent for Europe, Asia, and Africa, respectively; and 2.6 per cent for the whole world). This was strongly correlated with the changes these two continents made in their immigration policies. In the late nineteenth century, Australia, Canada, New Zealand and the United States adopted anti-Chinese immigration policies. Such discriminatory policies were not changed until the latter part of World War II in New Zealand and the middle of the 1960s in the other three countries. Especially since the 1980s, these historically immigrant countries have provided Chinese with more immigration opportunities through a series of legal and policy adjustments. In the case of the United States, a couple of policies have facilitated the inflow of Chinese population, including the Refuge Law of 1980, the yearly 20,000 immigration quota for both mainland China and Taiwan introduced in 1981, the Immigration Reform and Administration Law of 1986, and the Immigration Law of 1990. A historical development of the correlation between the immigration patterns and policies in the United States is presented in table 15.5. It clearly shows that changes in the size of the inflow of Chinese immigrants and of the overseas Chinese population have been significantly determined by the immigration policies adopted by the United States during the different periods from 1860 to 1990.

References

Chen, Da (1939), *Overseas Chinese in the Southern Asia and Societal Environment of Fujian and Guangdong* (in Chinese). Beijing: Commercial Press.

Chen, Zhexiang (1960), The contract system of Chinese labourers in the 19th century, *History Studies*, 63 (1) (in Chinese).

Espenshade, Thomas, J. and Wenzhen, Ye (1994), Differential fertility within an ethnic minority: the effect of 'trying harder' among Chinese–American women. *Social Problems*, 41 (1), 97–113.

Ford, K. (1990), Duration of residence in the United States and the fertility of US immigrations. *International Migration Review*, 24 (1), 34–68.

Hing, Bill Ong (1993), *Making and Remaking Asian American through Immigration Policy: 1850–1990*. Stanford: Stanford University Press.

Kahn, J. R. (1991), Immigrant and native fertility in the U.S. during the 1980s. Paper presented at the 1991 Annual Meeting of the Population Association of America, 21–3 March.

Lee, E. (1966), A theory of migration, *Demography*, 48 (3).

Li, Jinming (1986), The development of personal overseas trade and the high tide of international population flow. In Research Institute of Overseas Chinese at the University of Overseas Chinese (ed.) *Collection of the Research Papers on Overseas Chinese History*, 5 (1), 1–17.

Mark, D. M. L. and Chih, Ginger (1982), *A Place Called Chinese America*. The Organization of Chinese Americans, Inc.

Peng, Jiali (1980), Chinese labourers working in the Western colonial areas in the nineteenth century, *World History*, 80 (1), 3 (in Chinese).

Poston, D. L. Jr and Mei-Yu, Yu (1990), The distribution of overseas Chinese in the contemporary world, *International Migration Review*, 24 (3), 480–506.

Poston, D. L. Jr et al. (1995), The distribution of overseas Chinese in the contemporary world, *Population Sciences of China*, 95 (1), 9–15.

Stephen, E. H. and Bean, F. D. (1992), Assimilation, disruption, and the family of Mexican-origin women in the United States, *International Migration Review*, 26 (1), 67–88.

Thernstrom, Stephan (ed.) (1980), *Harvard Encyclopedia of American Ethnic Groups*. Cambridge: Harvard University Press.

Wang, Gungwu (1991), *China and the Chinese Overseas*. Singapore, Times Academic Press.

Wang, Xiuyin and Du, Fu (1992), *Studies of Population Migration* (in Chinese). Qingdao: Haiyang University Press.

Wu, Fengbin (1988), *The History of Contracted Chinese Labourers* (in Chinese). Nanchang: Jiangxi People's Press.

Ye, Wenzhen and Lin, Qingou (1994), An exploratory study of undocumented emigration from Fujian. *Overseas Chinese History Studies*, 95 (1) (in Chinese).

Ye, Wenzhen and Lin, Qingou (1996), A study of foreign-related marriages in Fujian Province of China, *Population and Economy*, 96 (2) (in Chinese).

Zhang, Shanyu (1992), Problems of population migration in China. *Population Research*, 92 (2), 16–22 (in Chinese).

Zheng, Ming et al. (1985), *Overseas Chinese* (in Chinese). Beijing: People's Press.

Zhu, Guohong (1987a), A historical observation of international migration in China. *Population Research*, 87 (4), 24–9 (in Chinese).

Zhu, Guohong (1987b), International migration in Fujian Province. *Population and Economy*, 87 (1), 38–41 (in Chinese).

16 The Ethnic Minority Population in China

Du Peng

China is a country made up of 56 nationalities. The Han have the largest population, the others being ethnic minorities. In 1995 there were 108.46 million members of ethnic minorities, constituting 8.98 per cent of China's total population. Minority peoples live in every province, autonomous region, and municipality together with the majority Han people, but are more concentrated in southwestern and northwestern China, as well as in most border areas, regions which in all cover about 60 per cent of China's total area. Except for the five autonomous regions of Inner Mongolia (Nei Menggu), Xinjiang, Tibet (Xizang), Guangxi and Ningxia, provinces such as Yunnan, Sichuan, Guizhou, Qinghai, Gansu etc. are the major areas accommodating minority populations (see (figure 16.1). This distribution pattern has taken shape throughout China's long history of development, as ethnic groups migrated and mingled (White Paper, 1998).

The minority regions have vast areas with sparse populations and rich natural resources. They provide the indispensable material foundation for China's attaining extensive development in industry, agriculture, forestry, animal husbandry and tourism. Nevertheless, for a long time, population data on each of the ethnic minorities had been lacking, and it was difficult to make an in-depth study and analysis of the minority populations. The 1990 population census collected data on the ethnic minorities, which enable us to analyse the characteristics and demographic dynamics of the various ethnic minorities, as well as the problems they have. The census data is the base for this chapter.

Minority Populations in the midst of Socio-Economic Development

The quantitative development of minority populations

Before the People's Republic of China was founded in 1949, the population pattern in most minority populations was a traditional one, characterized by high birth rates and mortality rates and a low natural growth rate. Their growth was slow: some of them had been at a standstill for a long

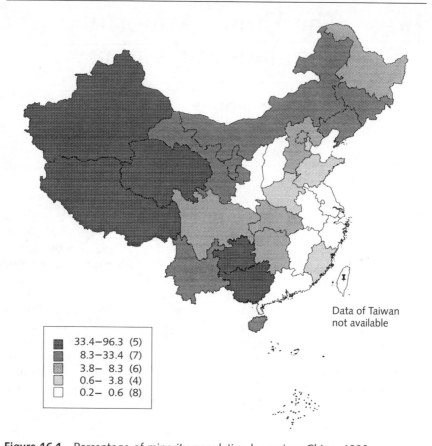

Data of Taiwan
not available

33.4–96.3 (5)
8.3–33.4 (7)
3.8– 8.3 (6)
0.6– 3.8 (4)
0.2– 0.6 (8)

Figure 16.1 Percentage of minority population by region, China, 1990

Source: China Population Information and Research Center, *China's 4th National Population Census Data Sheet.*

time, others had been gradually decreasing in number, a few were nearing extinction.

After 1949 the Chinese government adopted and enforced a set of policies for national equality, unity and common prosperity, aimed at energetically developing the economy, culture, education, medical and health services, and communications of the minority areas. Meanwhile, experts and scholars conducted large-scale investigations, held scientific discussions and distinguished between the ethnic minorities. These efforts created conditions for the population growth of the minorities, whose numbers increased gradually.

When China conducted its first census in 1953, it identified 38 ethnic minorities with a total population of 35.32 million. In 1964 the second census was carried out, which identified 51 ethnic minorities with a total population of 39.99 million. Then in 1982 the third census confirmed that there were 55 ethnic minorities totalling 67.30 million people. The fourth census took place in 1990, by which time the total population of the 55 ethnic minorities had grown to 91.32 million.

In the 11 years between the first and second censuses, the total population of China's ethnic minorities increased by 13.2 per cent while the national population grew by 20.1 per cent. Then in the 18 years between the second and third censuses, the population of the ethnic minorities jumped 68.3 per cent while the national population rose 42.2 per cent. In the 8 years between the third and fourth censuses, the population of the ethnic minorities increased by 35.2 per cent, while the national population increased by 12.5 per cent. The growth rate of the minority population was thus more than twice that of the national population. It was the period that recorded the fastest growth in the minority population, partly owing to the nationality identification policy carried out after 1982.[1] It is reported that between 1982 and 1987, about 10 million people originally identified as members of the Han population changed their nationality to a minority one. Its influence on population growth is conspicuous in the figures for the Manchu, Tujia and some other nationalities.

Today China has 18 ethnic minorities that have a population of more than one million each. There were only 10 such minorities in 1953 and 1964, and 15 in 1982. This shows that the population of China's ethnic minorities has passed from slow to rapid growth. The population growth of China's ethnic minorities has been facilitated by many factors, including changes in socio-economic conditions, the population policy and the policy of distinguishing between nationalities.

Fertility of ethnic minority women

The minority population in China enjoy special treatment in marital and reproductive issues. China's Marriage Law set the legal minimum age at marriage as 20 years old for women and 22 for men. While this law is implemented in the minority autonomous areas, the minimum ages are different: it is 18 for women and 20 for men. A similar situation prevails in the family-planning programme as well. A nationwide family-planning programme started in the early 1970s. The fertility rate of Han people dropped rapidly in the 1970s and 1980s. However, birth control regulations were more relaxed for the minority population. Hence ethnic minorities maintained a high fertility rate over a long period of time. Since the 1980s the State has established 'Controlling population growth while upgrading population quality' as its basic national policy. It has begun to advocate family planning among the ethnic minorities. Since then, the total fertility rate (TFR) of minority women has fallen notably; it dropped from 4.24 in 1981 to 2.91 in 1989, a decline of 31.4 per cent. The TFR gap between Han women and minority women narrowed significantly.

Sharp decline in minority population mortality rate

The State has adopted numerous measures to help ethnic minorities and minority areas develop their economies, cultures, and medical and health services. Institutions have been widely established in minority areas to train medical personnel to meet the shortage of doctors and medicines. This has laid a good groundwork for raising the physical quality of the population and curing various diseases. As a result, the mortality rate of the minorities has dropped remarkably within a short period of time.

In 1990 the crude death rate of China's population was 6.28 per thousand, that of the Han people being 6.22 per thousand. The ethnic minorities that reported a low death rate were the Xibes, Uzbeks, Mulams, Hezhens, Manchus and Yugurs, each of which had a crude death rate of less than 5 per thousand. Those ethnic minorities who registered a crude death rate of higher than 10 per thousand were the Lhobas, Vas, Drungs, Moinbas, Blangs, Benglongs, Lahus, Nus, Jingpos, Lisus, Hanis and Tajiks. There was a great difference in mortality rate between China's nationalities, the highest being more than twice as great as the lowest. The higher mortality level of ethnic minorities was equal to the national average in the first half of the 1970s, or to the average for developing countries. The low mortality rates of some Chinese ethnic minorities ranked among the lowest in the world.

In 1990 the life expectancy of China's total population was 69.7 years. Those nationalities who had a higher life expectancy than this were the Mulams, Xibes, Manchus, Huis, Hans, and Shes; they also reported a low mortality rate. With a higher mortality rate, the Vas, Lahus, Blangs, Nus, Jingpos, Hanis, Lisus, and Ewenkis had a life expectancy less than 60 years. The difference between the longest and shortest life expectancies was about 10 years.

Most of the nationalities that enjoy a low mortality level live on the plains or in the eastern part of China, where the economy is more developed and communications are more convenient. On the other hand, most of the nationalities with a high mortality rate live in southwestern China, especially in the valleys of the Lancang and Nujiang Rivers. The regional and national characteristics of the mortality difference, as well as the contributing factors, remain matters for in-depth study. It is certain, however, that there is a close relation between mortality rate and social development.

Rapid improvement in the educational level of the ethnic minorities

Before the founding of the People's Republic of China, the ethnic minorities maintained backward economies and had few educational opportunities. After the People's Republic of China was founded, the State enforced special policies to help the minorities develop their economies and education systems. Measures were taken to establish various kinds of national schools and colleges and to promote minority students. Since the 1980s the educational level of the minority population has been raised considerably. In 1990, 43 out of every 1,000 members of the Korean minority people had a university or college background, three times the national average. Out of every 1,000 people across China then, 14.2 people had had a university or college education. The national average was surpassed not only by the Koreans, but also by the Mongols, Manchus, and Huis. The adult illiteracy rate among all nationalities was lower in 1990 than in 1982. According to the census data, the Koreans and Manchus in 1990 had the lowest adult illiteracy rates, 7 per cent and 11.4 per cent, respectively. The other ethnic minorities had also improved their educational standing.

The Basic Characteristics of the Minority Population

Vast areal distribution, sparse populations and low density

The 1990 census found China's total minority population to be 91.32 million, accounting for 8.04 per cent of the national population. It was, however, mainly distributed over immense areas that together make up about 60 per cent of China's total land surface, though there were also minority residents in every city and county in the country. In 1990 the average density of the national population was 118 per square kilometre, while that of the minority autonomous regions was only 24.8 persons per square kilometre. The lowest population densities were found in Tibet (1.8 persons per square kilometre), Qinghai (6), Xinjiang (9), and Inner Mongolia (18), where ethnic minorities live in homogeneous communities.

Areas inhabited by many or several nationalities

Time and again, ancient Chinese rulers had stationed garrison troops on the frontiers to open up areas of wasteland and then moved people there. This led to mass migrations and exerted an impact on the concentration of ethnic minorities. From the national point of view, China is a huge family made up of various nationalities living together. In some specific areas, however, ethnic minorities are living in homogeneous communities, with one or two nationalities predominating. For instance, Xinjiang, Tibet, Ningxia, Guangxi and Inner Mongolia each have their own main ethnic minority population, justifying their establishment as autonomous regions. In addition, there are 30 minority autonomous prefectures and 122 minority autonomous counties. The communities in these autonomous areas are not so absolute. In them the main ethnic minorities are living together with other nationalities. The Huis are the most widely scattered ethnic minority in China. Besides their own autonomous areas, they are found in 97.3 per cent of China's cities and counties.

Minority population numbers

In 1995 ethnic minorities made up only 8.98 per cent of China's total population, but they numbered 108.46 million, which is equal to the total population of several medium-sized countries. Among China's numerous ethnic minorities, 18 had populations topping one million each: the Zhuang population came to 15.56 million, the Manchu population to 9.85 million, and the Hui population to 8.61 million.

Population sizes, however, vary widely between minorities. Among the 55 ethnic minorities, some have populations numbering over 10 million, some have fewer than 10,000 members. Most of the minorities have a population of under one million. In 1990 there were 30 minorities each having a population of less than 300,000. Seven of them had less than 10,000 each. The Lhobas had the smallest population – 2,312 people in 1990.

Young age structure

In 1982 China's minority populations basically had a young age structure, children made up 40 per cent of the population of 10 of the 15 minorities that had a population of more than one million. The median age of the population of 16 ethnic minorities was under 20 years. Since the 1980s, the ethnic minorities have made much headway in popularizing family planning, the birth rate has dropped, the age structure has been passing from the young model to the adult model. In 1990, of the 18 minorities with a population of over one million, only two had children making up 40 per cent of their population, and three had a median age of under 20 years. The age structure of Koreans, Manchus and Russians had changed to an adult model, but the great majority of the ethnic minorities still had a young age structure.

Imbalance in the population growth of ethnic minorities

There is imbalance in the population quantity of ethnic minorities and in their educational background and geographic distribution. The Zhuang have the largest population of 15.56 million and the Lhobas the smallest population of 2,312, a far cry from each other. The adult illiteracy rate of the Tibetans is as high as 69.4 per cent, while that of the Koreans, who have the highest educational standard in China, is as low as 7 per cent. The highest population density among the ethnic minorities is 178 people per square kilometre and the lowest 1.8 people.

Issues Relating to Minority Population Growth

As we have discussed briefly, the ethnic minorities are confronted with the rapid growth of their populations, which is mainly caused by the time lag in carrying out the family-planning programme.

Continued rapid growth in the near future

By 1990 the minority populations totalled 91.32 million, accounting for 8.04 per cent of China's total population; they had increased by 1.58 times since 1953. The increase was caused both by natural growth and by other factors. They have already passed the 100 million mark and will continue to grow in the near future if their current growth rate is maintained. The great majority of the ethnic minorities have a growth rate higher than the national average. While the total fertility of the Han is approaching the replacement level, it remains at around 4 for many minority groups. Consequently, ethnic minorities have a young age structure with the number of their childbearing-age women increasing and causing their populations to grow rapidly in the coming period of time.

Family planning and upgrading population quality

The broad masses of the ethnic minorities cherish the idea that population growth should tally with the development of the local economy and education and should fit in with the local natural environment and resources. From their own practice they are gradually learning that it is necessary to carry out family planning, because population overgrowth will not only keep them from upgrading their standard of living, but affect the development of their economy and education, their prosperity and progress. The national policy of controlling population quantity and upgrading population quality has gradually struck root in the hearts of ethnic minorities. They are taking voluntary action to develop their economy, culture, education, technology, and medical services as well as to upgrade their standard of living. However, population policy towards minorities will most likely remain flexible. Meanwhile, to providing a quality service of reproductive health to a sparsely settled minority population is a great challenge to China's family-planning programme.

Elevating the status of minority women

Influenced by feudal ideas, the status of China's women remained low for a long time, and large numbers of minority women suffered even more than Han women. Nevertheless, the government has made efforts over the past 40 years and more to change the situation. As a result, the status of the minority women has been raised considerably. From 1982 to 1990, the proportion of women receiving education in the minority populations increased greatly. In 1990, out of every 1,000 minority women, 521.5 had benefited from schooling. This proportion was lower than for Han women, who had 635.5 with education in every 1,000. But some ethnic minorities had a larger proportion of educated women in their population than the Han people. The Koreans had 794.9 in every 1,000, the Manchus 730, the

Kazaks 679.9 and the Mongols 653.4. Likewise, in 1990, employed minority women (aged 15 and over) across China totalled 22.486 million, amounting to 76.43 per cent of the age group, whereas the proportion of employed women nationally was 73.02 per cent of the age group. A higher educational level and larger proportion of employment have caused the minority women to play a daily increasing part in social life, raising their status. In addition, they are playing a greater role in bringing up fewer but healthier children.

Problems of Population, Resources and Development of the Ethnic Minorities

The areas inhabited by the ethnic minorities have plentiful natural resources. China ranks first in the world in water and coal resources, which are mostly distributed in minority areas. The minority autonomous regions make up 63.7 per cent of the total area of China, 94 per cent of national grasslands, and 41.6 per cent of the national forests. In detail, the forestry reserves of the minority regions constitute 51.1 per cent of the national total and their water resources 52.5 per cent. The per-capita resources of the minority regions are enormous.

Nevertheless, the minority regions are economically backward. Their resources have not been well tapped. Many minority populations suffer from poverty. Due to the restrictions and influences of historical, physical, geographical and other factors, central and western China, where most ethnic minority people live, lags far behind the eastern coastal areas in development. In some ethnic minority areas, the people are inadequately fed and clothed, while in some other areas sustained development has been adversely affected by poor production conditions. Among the 592 poverty-stricken counties listed in the national Poverty Alleviation Plan in 1994, 257 were minority counties, accounting for 43.4 per cent of the total number of poor counties and 40 per cent of the total number of counties in minority autonomous regions. The average income of the poor minority counties is lower than that of other poor areas in the country. Moreover, owing to harsh natural conditions, such as rocky areas in southwestern China, high and cold areas and dry and desert areas, efforts at poverty alleviation have been beset with difficulties, and it is easy for better-off people to be reduced to poverty again in these areas. At present, poverty still affects 50 million Chinese people. At least 20 million of them belong to ethnic groups (*People's Daily*, 22 January 1999).

More assistance is being promised by the Chinese government to help lift poor ethnic minority people out of poverty, Help will be provided first to ethnic minority groups who have relatively better living conditions and development options and lower populations. It is reported that between 1996 and 1998, the central government allocated 16.95 billion yuan, or 45 per cent of total poverty alleviation funds to the 257 poor minority counties. However, more preferential policy treatment and direct investment are needed to develop these areas. Otherwise, in the long run, the harmonious development of the national economy and economic returns will be affected.

The radical solution to poverty, however, depends on the ethnic minorities themselves, on reconstruction through their own efforts. The most pressing tasks in the minority areas are controlling population growth, upgrading population quality, speeding up the tapping of resources, and protecting the environment. In the support-the-poor drive, family planning should be integrated with curing poverty and illiteracy to provide the minorities with more opportunities for development as soon as possible, to eradicate their backwardness, and to achieve common prosperity for the Chinese nation as a whole.

Note

1. The policy allows people to choose or change their nationality if their parents or grandparents belong to different ethnic groups.

References

China State Statistics Bureau (1993), *Tabulation of the 1990 Population Census of the People's Republic of China.* Beijing: China Statistics Press.

Information Office of the State Council of the People's Republic of China (1999), *White Paper on National Minority Policy and its Practice in China.*

Wu, Cangping (ed.) (1997), *General Report of China's Changing Population and its Development.* Beijing: Higher Education Press.

17   Health and Health Care in Transition

Tang Shenglan and Wu Zhuochun[1]

China's population, once known as 'the Sick of East Asia', was among the least healthy in the world prior to the founding of the People's Republic in 1949. The burden of disease was both a consequence and a cause of the nation's poor economic performance (World Bank, 1996). From the 1950s to the 1970s, the health status of the Chinese population, particularly in the rural areas, improved significantly, while the national economy and household incomes remained low. This was true largely because health care had strong political backing, and basic health services were accessible to the vast majority of the population at an affordable price.

Since 1978 various reforms have had profound repercussions on the health care system and thus on health improvement. Using data from some empirical studies, this chapter discusses changes in the organization, financing and provision of health care and in health status in the urban and rural areas during the transitional period, and examines the challenges and opportunities China is facing in making health development sustainable.

Health Care Prior to Economic Reform

Health care in China was highly politicized during the period between the 1950s and 1970s. Health providers at all levels actively participated in public health campaigns, and responsibly undertook their tasks as part of their political commitment to the construction of a socialist country. By the mid-1970s China had established a highly structured health sector. In the rural areas, most villages (former brigades under the commune system) had a health station staffed by one of the well-known 'barefoot doctors', who were trained as part-time health workers. They led public health programmes and provided basic medical consultation. Most townships (former communes) had health centres that provided referral services and supervised preventive care. Each county had at least one general hospital and two specialized institutions that organized disease prevention and maternal and child health care. All health facilities and workers, under the leadership of the county party committee, worked together following the guidelines of a health development plan developed by national and/or provincial health authorities. Higher-level health facilities provided technical support and supervision to lower-level health facilities. Conversely, lower-level health facilities were willing to co-operate with the higher levels in the implementation of preventive programmes and in health information collection. In addition, the co-operative medical scheme (CMS)[2] implemented in most villages facilitated co-ordination and linkage among three levels of health facilities, while ensuring that most rural population were able to pay for care at an affordable price.

Urban health care was largely organized and developed by health authorities at the provincial and municipal levels under the leadership of the Ministry of Health. However, other industrial sectors, such as mining and telecommunication, also had their own health facilities that were mainly responsible for providing the employees with health services. Like the rural areas, there was a structured three-tier network of urban health care established in the period of 1950s and 1970s, that is, municipal/district hospitals, street hospitals and community health stations (*li-long-wei-sheng-shi*). Some large cities also had a number of teaching hospitals affiliated with local medical faculties at universities or colleges. Unlike the rural areas, however, the organization and delivery of health services, was not well planned and developed. People covered by the government insurance scheme (GIS) and the labour insurance scheme (LIS)[3] would usually have to seek care directly from the designated hospitals, though those uncovered by any insurance obviously had a right to visit any health facility they wished. Most street hospitals and community health stations were poorly staffed. Consequently, some street hospitals and community health stations were under-used. The referral system never worked well. Free medical care was provided by the GIS and the LIS to all employees and retirees, as well as to university and college students. Preventive and promotive care was provided jointly by the anti-epidemic stations (AES) and the maternal and child health centres (MCHC) at city and district levels, as well as by community health stations. Relatively speaking, the preventive and promotive care was highly cost-effective in the improvement of health status for the population.

Health Sector Reform – an Overview

Since the beginning of the economic reform, a number of radical reforms associated with the health sector have inevitably been introduced in the process of adapting to a market economy. In rural areas, the transition from agricultural collectives to a household responsibility system weakened the financial base of the CMS. As a result, in the mid-1980s only 5 per cent of villages were operating the CMS, as opposed to some 90 per cent in the late 1970s. Health services, especially in the poor areas, have been showing signs of declining, with fewer people able to pay for the services they need (Tang et al., 1994). In urban areas, the virtually free medical care provided by the GIS and the LIS has led to a rapid increase in medical care costs. Consequently, both governments at all levels and state- or collective-owned enterprises have been under great pressure to pay medical bills. Several measures, such as co-payment, have been taken by local governments and enterprises to control the rapid increase in medical care costs. More significant reform of the urban health care system has been experienced in some selected cities under the auspices of the State Council since the early 1990s. In the meantime, other reforms have also been introduced in the health sector. For example, the government has increased the autonomy of health facilities and encouraged them to rely on user fees, not state funding, to support their operations. The government health services have been also radically decentralized. All these changes have affected access to health care, the efficiency of service provision, and ultimately health status.

Rural areas

The impact of the economic reform on the organization of rural health services in China has been very significant. Decreased political motivation has affected the performance of health providers who used to co-operate fully with the government. Such full co-operation, however, no longer exists, since health providers have to protect their own interests. The changes in rural administrative structure and the organization of agricultural production have also had impact on the organization of rural health services including the operation of the CMS in most rural areas.

Village-level health services have changed greatly. The previous system of funding village health stations and paying health workers has disappeared. Many village health workers have to make a living by charging fees to patients, selling drugs for profit, and spending time on non-health activities. Not only did the number of health workers fall from 5.5 million in 1978 to 1.7 million in 1988,[4] but most village health stations were privatized and they have been acting as private practitioners (Tang et al., 1994). The statistics published by the MoH (1996) show that 14 per cent of villages in 1990 had no health worker. Such a phenomenon is very common, particularly in the poor areas. Tibet, Guizhou, Hainan, and Zhejiang were the provinces where 20–70 per cent of the villages had not set up health stations. That does not mean necessarily that all people in these areas are not able to get convenient access to basic health services, but it cannot be denied that a substantial number of the rural population in some poor areas have to travel a long way to seek medical care.

Township-level health services have also changed a great deal, albeit differently in the rich and the poor areas. In the rich areas some have expanded their services and acquired new technologies in order to stimulate utilization (Xiang and Hillier, 1995). Health centres in the poor areas have faced severe difficulties and some of them have had to close (Hillier and Xiang, 1991; Liu et al., 1996). The number of the township health centres declined significantly, from some 55,000 in 1978 to 48,140 in 1991. Most closures happened in the poor areas, owing largely to lack of funding and skilled personnel.

At the county level, county general hospitals have been granted more autonomy in the organization of health services, while being encouraged by local governments to generate more revenue from user fees to cover operating costs. The preventive institutions at the county level, such as AES and MCHC, have been allowed to provide curative services and sell drugs to local people, while being requested to continue to carry out their mandates in relation to the implementation of preventive care programmes. The changes have in various degrees jeopardized the preventive care programmes, particularly in the poor areas, although in some rich areas the preventive care programmes have been strengthened because of increased revenues from user fees.

The planning and co-ordination of health services has been weakening, owing to the diminished political control and the decentralization of administrative and financial responsibility for the health facilities. Along with increased managerial autonomy and the increasing importance of revenue generation via service fees in the health facilities, the capacity of the government to influence their behaviour directly has diminished significantly. Although health-development plans have still been developed on a short- or long-term basis at various levels, they have been less likely to be implemented completely. Most health providers would be in favour of carrying out some tasks that could lead to revenue generation. Co-ordination has also become more and more difficult as service providers compete with each other. For example, the health workers at the AES and the MCHC have less time to co-ordinate with village and township health workers in the preventive programmes, simply because they want to devote more time to providing local people with services that can increase revenue.

Urban areas

The organization of urban health services has also been affected by a series of economic and health sector reforms. Among the major changes are the operation and performance of health facilities, the development of a community-based health care, and the reform of work-related health insurance schemes (i.e. GIS and LIS).

As the health services have been decentralized and the government has contributed a smaller portion of health expenditure to almost all the health facilities, health managers have had an increased autonomy to organize provision of the services and manage the facilities. Most services provided have become profit-driven, and might not be cost-effective. The problem has also been associated with the financing mechanism and the provider payment system. Various measures, such as contracts between health authorities and providers, and between hospital directors and heads of hospital department, have been taken to provide health providers with a financial incentive, among other things. As a consequence, the performance of the health facilities has been changing as they adapt to competition.

Given a rapid increase of medical care costs and an increased burden of chronic non-communicable diseases, both the central government and some local governments have developed new approaches aimed at strengthening community health care in order to provide urban residents with easy access to essential clinic services and preventive care, and to control medical care costs. Home visits by physicians and home beds managed by tertiary hospitals are emerging in some cities. Some local governments have also considered the introduction of a community-based health care model being used in many western countries (e.g. GP in the UK, family physician in Canada and the USA). These initiatives are welcomed by people, on the whole, but face the challenge of how they should be accommodated in the current health-care financing system.

The hottest issue was how to transform the GIS and the LIS into an appropriate and sustainable health insurance schemes. There were a number of initiatives developed by local governments and health authorities during the 1980s, however, none of them have had significant influence at national level. Since the late 1980s and early 1990, Shanghai, Hainan and Shenzheng, among others, have implemented some radical reforms with different characteristics, tackling the problems of cost escalation, efficiency and equity (Zuo et al., 1998). Two cities – Zhengjiang and Jiujiang – under the sponsorship of the State Council, launched a demonstration health insurance project in December 1994. Finance for the health insurance comes from payments from enterprises or public agencies (10 per cent of employees' wages) and employees as well (1 per cent of their wages). The funds collected are committed to individual and group accounts (around 55 per cent and 45 per cent, respectively). People are required to use their personal medical accounts first in seeking medical care, and then have to pay for care out of their own pockets up to 5 per cent of annual income, after they run out of personal medical savings. Finally, a social risk-pooling fund, from the group account, covers a large proportion of medical care costs. The personal medical savings accounts plus large co-payments should encourage moderation of patient demand yet provide stop-loss coverage to protect against catastrophic medical bills. The Zhenjiang model includes two additional innovations. First, the formerly separate GIS and LIS have been combined into a single insurance scheme. Second, package fees are set per outpatient visit and per inpatient admission. The payment rates are established prospectively. Results from the first two years of operation appear to be positive, but a systematic and comprehensive evaluation is needed to confirm the results.

There have been quite a number of changes in the health sector emanating from the economic reform being undertaken in such an increasingly decentralized society. Some changes have brought health services closer to the needs of the population. But others have had negative impacts on access to health care and efficiency of service provision.

The Financing and Provision of Health Services

Prior to the early 1980s, there were three major sources of finance to cover to a large extent medical care expenses for the population. In the rural areas, the CMS funds contributed jointly by local collectives and individuals covered most of the expenses of health care for the participants. In the urban areas, the government via the GIS, and the state- and collective-owned enterprises via the LIS, paid for the medical care of their employees and retirees. In addition, almost all the state-owned enterprises, as well as some collective-owned enterprises, were responsible for 50 per cent of medical care expenses for direct dependants of their employees (called half LIS in China – *ban-lao-bao*). A small portion of the urban population was left to pay all their medical care costs out of their own pockets. In addition, the government financially subsidized government health facilities by paying salaries and funding several preventive programmes.

There were over 7,000 general hospitals and some 5,000 preventive and promotive health facilities built up at the county and higher levels by the end of 1970s, providing the urban population with both curative and preventive services. More western doctors than Traditional Chinese Medicine (TCM) doctors were trained, making the proportion of the TCM doctors decline from three quarters of all doctors in 1949 to one fifth in 1988.

The economic reform has brought about significant changes in the sources of health finance (World Bank, 1996). Contributions from the governments at various levels to the total national health expenditure reduced from 28 per cent to 14 per cent. The share of the CMS fell from 20 per cent to 2 per cent. On the other hand, out-of-pocket payment rose from 20 per cent to 42 per cent over the same period. Such a transformation in the financing base of the health sector has inevitably had implications for the provision of health services. The rest of the section looks at changes in financing and provision of health services in rural and urban areas separately.

Rural areas

Since the economic reform, the financing and provision of rural health services has shown a mixed picture. In some rich areas such as Jiangsu province, the rural health services has been greatly strengthened, however, financing and provision of rural health services in most poor areas has been in jeopardy. A large-scale national health service survey conducted in 1993 produced a number of key findings on the financing and provision of health services.

In the period between 1986 and 1992, there was a significant increase in health expenditure in the rural areas, most of which was spent on curative services. The AES, the MCHC and other preventive health facilities were not given favourable treatment in resource allocation by many local governments. The increase in average health expenditure per capita was much higher in the urban areas than in the rural areas. In 1992 about 10 per cent of the rural population were covered by the CMS and 85 per cent had to pay for medical care out of pocket. Approximately 70 per cent of outpatient services were provided by village health stations and township health centres, 11 per cent of the services were provided by county hospitals, and 15 per cent by private practitioners and clinics. In the rural areas there were twice as many patients admitted to the township health centres (18.55 million) than to the county hospitals (8.35 million). Provision of preventive and promotive services, particularly in the poor areas, was far from satisfactory. Only one quarter of the women living in the poorest areas got access to ante- and post-natal care.[5] Even in the richest areas one quarter of women did not seek care for various reasons. It was also found that the immunization coverage in the rural areas, particularly in the poor provinces (e.g. Shanxi and Guizhou), was far below the national average level.

One factor that is very much associated with outcomes in the health area is the introduction of fiscal decentralization (also called the financial responsibility system) by the government under the overall strategy for transforming its centrally planned economy towards a market one. It means that the lowest levels of government are responsible for many public services, while the central government allows the local governments to retain some taxes based on revenue-sharing contracts.[6] Thus, a growing proportion of revenues are retained and spent in the localities where the revenues are collected, indicating a tendency towards the regionalization of fiscal resources. Inter-regional inequality in financing of public health has substantially arisen.

The report from a national survey[7] shows that in the rich areas, local governments are able to increase the size of grants to health facilities. In the meantime, the health facilities are better able to rely on user fees to generate revenue. As a whole, the rural health services in these areas have been expanding with better equipment and new buildings. In contrast, many county and township governments in the poor areas have faced serious financial difficulties in providing services, although they continue to receive a lump sum fiscal transfer from the central government.[8] Medical equipment and buildings, particularly at the township level, are even worse now than two decades ago, since there has been no money for maintenance (World Bank, 1992). Preventive health-care programmes used to be fully funded by the governments. They are now in jeopardy because most government funding for preventive services barely pays the salaries of health

Table 17.1 Percentage of people who were not admitted to hospital in spite of referral by a doctor, 1988

	Rich counties	Intermediate counties	Poor counties
People referred but not admitted (per 1,000)	4	8	18
Admission rate (per 1,000)	60	28	24
Referral rate (per 1,000)*	64	36	42
Non-admission rate %**	6	23	42

Notes:
*Number of admissions plus the number of people referred but not admitted.
** Percentage of people who were referred but not admitted to the total number of referrals.
Source: Tang et al. (1994), *Financing health services in China: adapting to economic reform.* Research Report 26, Institute of Development Studies, UK.

workers, leaving little money for operating the programmes (Tang and Gu, 1996). As a whole, decentralization inhibits special assistance for the poor, since they generally live in areas with less capacity to tax and redistribute benefits through publicly subsidized health services.

Another factor that is crucial to the financing and provision of rural health services in China is the CMS. After the implementation of the household responsibility system in the early 1980s, which has weakened rural collective economies in general, together with diminished political backing and poor management, the CMS collapsed in most rural areas. Most rural populations again have to pay for medical care out of pocket. Table 17.1 sheds light on how many difficulties the rural population, particularly in the poor areas, faced in seeking hospital care in the late 1980s. The survey report shows that the poor spend a higher proportion of their income on health care than the rest of the population. Rich people are more likely to use health services of high quality while poor tend to use services of low quality. More strikingly, about 95 per cent of low-income service users who were referred to hospitals by doctors did not receive any services simply because they were unable to afford hospital care (Gu et al., 1995).

Urban areas

Government grants to urban health facilities have become a smaller proportion of total expenditure, accounting for less than 10 per cent in some large hospitals. Not only do the hospitals rely on user fees to generate revenues that are used to cover a substantial proportion of operating costs, purchase medical equipment and construct buildings, but preventive health facilities also have to charge fees for certain services they provide to supplement government grants. This has contributed to the rapid escalation of health expenditure in China.

About three-quarters of the urban population are fully or partially covered by the GIS and the LIS at present (MoH, 1994). However, those working in loss-making enterprises are less likely to get their medical bills paid by their employers. In the meantime, those working in profit-making enterprises enjoy a high quality of health services, since more high-technology equipment has become available. An effort made to reform the GIS and the LIS, which was described in the previous section, aimed to control the rapid growth of health expenditure, and redistribute health resources between rich and poor and between healthy and sick.

In the urban areas there is a very wide distribution of clinics and other health services at local level that are widely used by those who have a need for them (Henderson et al., 1994). According to the national health services survey in 1993, one-third of the urban population use the community health stations and the street hospitals as their first contacts for seeking care, while two thirds go directly to local district, municipal and other tertiary hospitals (MoH, 1994). Many large hospitals in China have invested a huge amount of resources to deal with simple medical consultations that are in general handled by GPs and family physicians in western countries. In addition, the provision of health services to an increasingly large floating population, who are not yet regarded as urban residents by most municipal governments, is an emerging issue for urban health planning. Some preliminary research findings show that the pregnant women among the floating population in Shanghai had difficulty in getting access to antenatal care or no knowledge that they were able to seek care.[9] There is, therefore, an advantage in establishing a community-based health care system in the urban areas to facilitate cost containment and access to basic health care.

The problems associated with the financing of urban health services are attributable partly to inappropriate mechanisms in the provider payment system, that is, fee-for-service, and in the financing of health facilities (i.e. profits from drug sales as a major source), and partly to changes in the financing of local government and in the operation of state-owned enterprises. Heavy reliance on fee-for-service provider payment methods is making GIS and LIS simply unsustainable. Meanwhile, allowing health providers to make a profit from drug sales has encouraged over-prescription, which has implications for service quality and cost (Zhan et al., 1998). Some price distortions (e.g. CT and MRI scan tests) have led to a misallocation of spending, medically inappropriate services, and upward pressure on overall health expenditure (World Bank, 1996). Hence, provision of health services in most health facilities is to a considerable extent profit-driven. The income of health workers is often linked to their capacity for revenue generation rather than to their provision of quality services.

Over the past decades, the focus was on the expansion of the health service, and, indeed, expansion was imperative, especially in the rural areas. In recent years there has been a growing interest in improving the quality of health care, an issue that is very complicated and challenging. In the urban areas, the quality of some services may have been improved, since more qualified doctors have been recruited and better equipment provided. However, over-prescription and inappropriate use of high medical technology have to some extent comprised the quality of the services. In the meantime, a lack of skilled health personnel, among others, is one of key problems resulting in lower quality of services in some poor rural areas. A study done by Gong and his associates (1997) revealed that the policy of allocating jobs by the state has been relaxed since the economic reform. As a consequence, more than 80 per cent of qualified doctors have been lost to poor counties during the last 15 years. The problem was found in quite a number of provinces including Hubei, Guizhou, Sichuan, Tibet, Shanxi and others.

Health in Transition – Progress and Problems

As mentioned in the introductory section of this chapter, the overall health status of the Chinese population has improved enormously since 1949. China's life expectancy in 1991 was 69 years, higher than the average level of mid-income countries, according to the World Bank's report (1993). The infant mortality rate (IMR) was also significantly reduced from over 200 per 1,000 live births in the 1940s to below 40 per 1,000 in the late 1980s (Gu, 1992). There are also a number of other indicators, such as incidence of infectious diseases, child growth and nutrition, that have shown a rapid improvement of the health status, China's population is ageing and heart disease, cancer and strokes have replaced malnutrition and infectious diseases as the major courses of death in the urban and some rich areas. On the whole, remarkable progress has been made in the health transition of the population over the past decades. One vital factor behind the success in health improvement in China, as has been widely acknowledged, is economic development and poverty alleviation. More important is, however, the health policy centred on disease prevention and community health, which influenced the way international organizations developed the strategy of primary health care in the late 1970s.

While the international health community and other developing countries have begun to learn from the experience of health development in China, new concerns about health problems have emerged, mainly focusing on two aspects: first, increased inequality in health status, particularly in children and women, between the urban and rural areas and between regions; second, an epidemic of non-communicable disease (NCD) resulting from environmental pollution, unhealthy life styles (smoking, alcohol, lack of physical exercise, etc), and demographic transition.

Increased inequality in health status

The problems mentioned above have undoubtedly had a negative impact on improving health, and particularly, narrowing the gap between the rich and the poor areas.

Table 17.2 shows a clear disparity in health status among the rural areas with different income levels and the urban areas in 1993. The higher the income level of the area was, the higher the life expectancy and the lower the infant mortality. Not only was there a significant differential in health status between the urban and rural areas, there was also an apparent gap between different rural areas. The World Bank report on China's health sector published in 1996 argued that China's earlier progress in improving child health appeared, in the aggregate, to have come to a stop, despite rapid income growth. The argument was based on the research findings showing that the under-five mortality rate had not declined since the mid 1980s. Some academics from China or Western countries might not fully agree with the conclusion, however, which was only based on one indicator. It may not be appropriate to rush to make a firm conclusion on the issue with limited evidence available. Nevertheless, one point that seems clear is that health improvement has not been equitable between the areas with different income levels since the economic reform. The results from a study on child growth in urban and rural areas of China done by Shen and his associates (1996) strongly supported this point. They found that in 1975 the average height of children in periurban rural areas was about 3.5 cm less than that of children in the urban areas. Between 1987 and 1992, the average height of both urban and rural children increased, but the net increase for rural children was only one-fifth that for urban children. They concluded that there were increased differences in height between rural and urban children and increased disparities within rural areas, in spite of an overall improvement in child growth over the past decades.

Table 17.2 Health status of the Chinese population in urban and rural areas

	Urban	Rural I	Rural II	Rural III	Rural IV
Average per capita income (yuan/year)	1,789	927	677	561	441
Infant mortality rate/1000	14	29	34	44	72
Life expectancy (years)	72	71	69	68	64

Source: The National Health Service Survey in 1993 and the 1990 Population Census.

A paper on gender and health recently published by Social Sciences and Medicine has also shed light on gender inequality of health status in the past decades (Yu and Sarri, 1997). The authors used the Physical Quality of Life Index (PQLI) and the Gender-Related Development Index (GDI) to examine changes in women's health status over the past five decades. They found that the overall level of physical well-being of Chinese women has increased, even in recent years, but the disparity in health between men and women, and between women in different areas (urban and rural, and rich rural and poor rural) is considerable. The GDI further indicates that China has seen significant progress in women's health, but has achieved far less with respect to gender equality overall. It is also revealed by the paper that the female IMR has been much higher than the male one for many years and there have been no signs that the gap can be narrowed down over the years to come.

Increased burden of chronic and non-communicable diseases

Urbanization and industrialization, rising incomes, the expansion of education, and improved medical and public health technology have resulted in a remarkable demographic and epidemiological transition. Fertility and infectious disease mortality have declined while chronic and non-communicable disease mortality has increased in many developing countries including China (Jamison et al., 1993). The health and disease pattern is therefore changing in China, as in other countries. Strokes, cancer, ischaemic disease, and chronic respiratory tract diseases account for most of the mortality among Chinese people of late-middle and older ages, although children still die from a relative short list of infections, and the incident rates for STDs and HIV/AIDs have been rising sharply over the past decade. Chronic and non-communicable diseases have increased their share of the disease burden in China, while infectious diseases continue their steady decline in relative importance. In addition, poisoning, injury and suicides in the rural areas are also been recognized as among the major causes of death.

The burden of these diseases in China is expected to increase dramatically in the next decades, because many potential risk factors have been greatly affecting the health of the population. Table 17.4 shows the age-standardized incident rates of hypertension and stroke in selected urban and rural areas of China, taken from a survey of a 100,000 population recently. The increased high prevalence of hypertension in China has been one of the major factors leading to strokes. Based on the statistics reported by the MoH (1996) the prevalence of hypertension in the population aged over 15 years in the urban and rural areas in 1991 was, respectively, 13.8 per cent and 9.4 per cent. The figures are most likely to rise, owing to an increasingly unhealthy diet and life style. According to the results presented in the table, Chinese people living in the north of the country have a higher chance

Table 17.3 Age-standardized incident rates of hypertension and stroke among the population aged 25–74 in 1991–1994

	Region	Hypertension (%)		Stroke per 100,000	
		Male	Female	Male	Female
North	Harbin/urban	45	26	595	282
	Beijing/urban	33	27	378	288
	Beijing/rural	40	28	303	139
	Hebei/urban	23	24	474	173
	Shanxi/rural	25	38	350	265
	Jiangsu/rural	17	16	172	107
	Shanghai/urban	16	15	122	86
	Sichuan/urban	14	12	126	69
	Guangxi/rural	16	16	232	159
	Guangdong/urban	17	17	153	119
South	Guangdong /rural	16	5	127	45

Source: Zhou B. F. (1996), Trend of cerebrocardiac disease in China in the early 1990s, *Chinese Journal of Prevention and Control of Chronic and Non-Communicable Diseases*, 4 (4), 145–9.

of suffering from hypertension and thus stroke than their counterparts in the south, on the whole. The possible factors behind the result might be that the northerners are more likely to eat fat and salted food, be smokers, and do less physical exercise because of cold weather.

Smoking is another big problem that can result in a significantly increased incidence of many chronic diseases. As the World Bank reported in 1996, by 1990 tobacco use accounted for about 800,000 of China's 8.9 million annual deaths. According to Murray and Lopez's projections (1996), there will be over 2 million deaths related to tobacco use in 2,020 and the percentage of tobacco-attributable deaths, as a proportion of all deaths, will almost double, unless tobacco use can be curtailed.

On the whole, therefore, a health transition has taken place in China that has brought about changes in the levels and causes of illness and death similar to those that have already happened in the developed world. New approaches will have to be developed and adopted to tackle the rising health problems. However, it is imperative for health policy-makers to bear in mind that the first Chinese health-care revolution – extending successful programmes for improving child health and control of endemic infectious diseases in the poor rural areas – has not yet been completed.

Making Health Development Sustainable – Challenges and Opportunities

As presented above, China had an excellent record in improving the health status of its people in the period from 1950s to 1980s. Since then, it has been facing several challenges in the reorganization and financing of health services to ensure that the vast of majority of the population are given equal access to health care, that the improvement in health status, particularly in the poor rural areas, can continue, that service provision by health facilities is adequately efficient and that cost containment, particularly in the urban areas, can be better achieved. All these are the key to an attempt to narrow the gap in health status between different regions (urban vs. rural, rich vs. poor) and to make health development in China sustainable.

One big challenge is how to reorganize health services in both rural and urban areas in adapting to the changing socio-economic environment. In the rural areas, strengthening the three-tier network of health care should be placed at a high priority by local governments. Whatever model of managing rural health care is adopted, local government will have to make sure that the public health facilities are appropriately staffed and equipped, as well as giving incentives to provide efficient and quality services, and that the private health sector is well regulated and monitored. Competition between the public and private health sectors should be encouraged. In the meantime, the government bears the responsibility for developing the CMS to help the rural population get access to basic health care. In some areas, local Chinese political leaders have already successfully mobilized people to participate in the CMS (Tang et al., 1994). In addition, the organization of preventive care programme should continue to be an important task of public health facilities at the county and township levels where local governments have an obligation to provide support.

In the urban areas, the reorganization of health services is under way or under consideration throughout the country. While the work-related health insurance schemes are being reformed in some cities, a new approach to the organization of urban health care, with an emphasis of community care, is also being developed, aiming to providing more accessible, affordable and effective services. That implies that the government needs to consider training and retraining the health professionals who are required to play the role of family physician, and developing mechanisms for organizing and managing the new service delivery system (i.e. home beds, referral systems, etc.). Additionally, the increasing floating population in urban areas, who have so far been neglected by the current system, will have to be given access to curative services and preventive care.

The second big challenge is how to finance health services within the context of a more liberalized economy. For example, the government has advocated in a recent policy document (State Council, 1997) that the CMS should be mainly financed by individuals with the support of local collectives and the government. The principle regarding sources of finance for the CMS has been clearly stated, but what has not yet been made clear is to what extent the local collectives and the government should support the CMS. Households are more likely to make financial contributions and join the schemes if they can expect to receive services worth more than what they have contributed. Hence it is very important for local collectives and the government to support the CMS adequately. In the meantime, the financing of preventive services should always be placed as the highest priority, since most preventive services have high cost-effectiveness. The government should make sure that the preventive care programmes, particularly in the poor areas, are adequately funded, if necessary, via special fiscal transfers. Only if the government is determined to do so, can the increased inequality of health status between the rich and poor areas be reduced to a minimum.

Financing of urban health services has been under increased pressure, especially since the beginning of the reform of the state-owned enterprises. The average cost of the GIS and the LIS rose 7.4 per cent a year between 1985 and 1989, after adjusting for inflation and the ageing of beneficiaries; and the cost has continued to increase (World Bank, 1996; Liu and Hsiao, 1995). Many state-owned enterprises have been, however, loss-making for a long time, and found it difficult to pay for medical care for their employees. Who (the central government, local governments, loss-making enterprises, or employees themselves) should take responsibility for paying for health insurance has become a critical issue in developing the new work-related health insurance scheme. Establishment of a safety-net is extremely important in the current situation in which more enterprises are expected to go bankrupt and more workers are going to lose their jobs in the coming years. Thus, the inequality of health status between different income groups should be minimized.

Given the limited health resources and the increased burden of chronic and non-communicable diseases, cost containment and efficiency improvements are also serious challenges, among others, in running the health services in China. It should be recognized that increases in health expenditure will not necessarily bring about improved health outcomes, if the investment in health is not wise. The current health care system has been fragmented and the performance of health providers has been not satisfactory in terms of cost containment and efficiency improvement. Health legislation has, since the mid 1980s, been introduced as a policy tool to ensure that some essential health services with reasonably good quality are accessible to the population, and that the performance of health providers is properly monitored and regulated. However, some health legislation has not yet been enforceable, particularly in the poor areas, due to a number of factors including financing and technical capacity (Tang, 1997). It seems that more strict and practical regulation and monitoring of health services should be developed by the government to reverse the current trends. Managed health care approaches will have to be used to improve the performance of all the health providers so as to increase value for money.

While China is facing many challenges in sustaining health development, it has also opportunities to achieve its goal of 'Health for All'. In November 1996, the Central Committee of the Communist Party of China and the State Council sponsored a high-profile National Conference on Health Reform and Development, as a serious response to these emerging problems. A Resolution on Health Reform and Development published in January 1997 affirms that the government and the Communist Party must play an active role in health development, via the formulation of appropriate policies related to health sector reform, and that the implementation of primary health care in the rural areas, the strengthening of preventive care programmes, and the reform of urban health insurance schemes should be put as the top priorities (MoH, 1997).

This provides a unique opportunity for China to meet its challenges and tackle its problems in the near future. China's history over five decades tells us that political backing is vital in health development. Developing appropriate health policies to address these problems has now become the most important step towards success in health development in the next century. China's economy, which has been developing fast for the past two decades, is also providing a sound financial base to overcome obstacles resulting partly from the previous planned economy and partly from the process of economic restructuring. In addition, many lessons from the reform of OECD health systems and others (i.e. Singapore) can be learnt by China in undertaking health sector reform. These include extending coverage and risk-pooling arrangements, cost containment and efficiency improvements. As a whole, the on-going reforms in the health sector will have to have clear objectives for improving equity in access to health care and efficiency in the use of limited health resources, and for developing effective means to achieve the expected goal.

Notes

1. The authors would like to express their sincere appreciation to Professor Peng Xizhe for his insightful comments on the early draft. They also wish to thank Mr Youde Guo of the Institute of Population Research of Fudan University for his assistance in preparing the figures used in this chapter.
2. Community-based pre-payment schemes, which are often financed jointly by township and village collectives and individual households, entitle members to reimbursement of a portion of medical expenses at local health-care facilities. Types of treatment covered and levels of reimbursement vary between localities.
3. The concept of traditional GIS and LIS differed from the one used in western countries. There were no third-party payers. Either the government or enter-

prises paid health providers on a fee-for-service basis for medical care of their employees.

4. The decline was partly attributed to people failing to pass the examination to promote barefoot doctors to rural doctors, and partly to loss of some of the barefoot doctors who dropped out of the health sector for various reasons.

5. The poor areas are the counties where the average per capita income per year was 441 yuan in 1992.

6. The contract-based fiscal system was changed to a rule-based system in January 1994, aiming to strengthen the centre's position in the collection of tax revenues (Zuo, 1994).

7. The survey was conducted in 20 counties. Counties with different income levels were included (Tang et al., 1994).

8. Only about 4 per cent of the total recurrent health budget in 1993 fell under the direct control of the central government (Berman et al., 1995). Province, county and township spending accounted for the remainder.

9. Personal communication with Prof. Shao-Kang Zhan of Shanghai Medical University in December 1997.

References

Berman, P. et al. (1995), *China's National Health Accounts*. Washington DC: World Bank.

Gong, Y., Wilkes, A. and Bloom, G. (1997), Health human resource development in rural China, *Health Policy and Planning*, 12 (4), 320–8.

Gu, X. (1992), Development of Chinese people's health status over 40 years, *Chinese Journal of Health Statistics*, 9 (4), 5–8 (in Chinese).

Hillier, S. and Xiang, Z. (1991), Township hospitals, village clinics, and health workers in Jiangxi Province: a aurvey, *China Information*, 6 (2), 51–61.

Jamison, D. T. et al. (1993), *Disease Control Priorities in Developing Countries*. Oxford: Oxford University Press.

Liu, X., Xu, L. and Wang, S. (1996), Reforming China's 50,000 township hospitals – effectiveness, challenges and opportunities, *Health Policy*, 38, 13–29.

Ministry of Health (MoH) (1991), *Selected Edition of Health Statistics of China* (in Chinese). Beijing, China.

Ministry of Health (MoH) (1994), *Research on National Health Services – An Analysis Report of the National Health Services Survey in 1993* (in Chinese). Beijing, China.

Ministry of Health (MoH) (1996a), *Chinese Health Statistical Digest* (in Chinese). Beijing, China.

Ministry of Health (MoH) (1996b), *Selected Edition of Health Statistics of China 1991–1995* (in Chinese). Bejing, China.

Murray, C. and Lopez, A. D. (eds) (1996), *The Global Burden of Disease and Injury Series. Vol 1*. Harvard School of Public Health, Cambridge. MA: Harvard University Press.

Shen, T. et al. (1996), Effect of economic reforms on child growth in urban and rural areas of China, *The New England Journal of Medicine*, 335, 400–6.

State Council of China (1997), *Decision of the Central Committee of the Chinese Communist Party and the State Council on Health Reform and Development* (in Chinese).

Tang, S. (1997), *Infectious Disease Prevention and Treatment in Poor Rural China – Legislation and Practice*. A Presentation at the Seventh Annual Meeting of the International Health Policy Programme, Arusha, Tanzania, 7–11 October.

Tang, S. et al. (1994), Financing health services in China: Adapting to economic reform, *IDS Research Report*, No. 26, IDS, Brighton, UK.

Tang, S. and Gu, X. (1996), Bringing basic health care to the rural poor, *World Health Forum*, 17 (4), 404–8.

World Bank (1992), *China: Strategies for Reducing Poverty in the 1990s*. Washington DC.

World Bank (1993), *World Development Report – Investing in Health*. Washington DC.

World Bank (1996), *China: Issues and Options in Health Financing*. Washington DC.

Xiang and Hillier (1995), The reform of the Chinese health care system: county level changes: the Jiangxi study, *Social Sciences and Medicine*, 41 (8), 1057–64.

Yu, M. and Sarri, R. (1997), Women's health status and gender inequality in China, *Social Sciences and Medicine*, 45, 1885–98.

Zhan, S. et al. (1998), Drug prescribing in rural health facilities in China: implications for service quality and cost, *Tropical Doctor*, 28, 42–8.

Zhou, B. F. (1996), Trend of cerebrocardiac disease in China in the early 1990s, *Chinese Journal of Prevention and Control of Chronic and Non-Communicable Diseases*, 4 (4), 145–9 (in Chinese).

Zuo, X. (1997), China's fiscal decentralization and financing local services in poor townships, *IDS Bulletin*, 28 (1), 81–91, Brighton, UK.

Zuo, X. et al. (1998), Reform of medical insurance system in the urban China: establishing cost control mechanisms and organizational innovation. Unpublished Research Report.

18 China's Environment and Environment Protection

Dai Xingyi and Peng Xizhe

Heavy Historical Ecological Debt

With a total territory of 9.6 million square kilometres, China is among the largest countries in the world. China is abundant in many natural resources, but its natural environment is not so friendly to the Chinese people. The plain in China only occupies 12 per cent of the country's total territory and most of the land surface is covered by mountains, hills, desert and plateau. These areas are prone to suffer from soil erosion, ecological deterioration and desertification under high population pressure and inadequate technologies. About 28 per cent of the land is almost useless for development with existing technologies, creating even greater population pressure on the other parts.

Relative to its population size, China has modest water resources, with an annual run-off flow of approximately 2,700 billion cubic metres. The major problem is a wide regional and seasonal variation in water availability. The main rainfall occurs in summer, often in the form of storms. In northern China the rainfall from June to September accounts for 85 per cent of the annual total. There is also a huge variation in rainfall between different years in some regions. Typically, in a drought year, rainfall may be less than one-tenth of what it is in normal years. Often there are several drought years in succession. Furthermore, there is great variation in water resources between areas. South of the Yangtze, the annual run-off accounts for 82 per cent of the run-off for the whole country, but the area has only 38 per cent of the farming land (see table 18.1). About 45 per cent of the territory is drought-prone with an annual rainfall of less than 200 mm. It is thus difficult and costly to develop water resources in China.

China is well endowed with mineral resources and can be self-sufficient in half of the mineral types it requires. However, resource per capita rates are very low. Furthermore, low-grade minerals unfortunately make up most of the reserves. For example, low-grade iron ore accounts for more than 98 per cent of total iron reserves, the average grade being lower than 34 per cent. Low-grade minerals require more energy in production and therefore

Table 18.1 Comparison of water resources between south and north China

	South China	North China
Cultivated land (%)	38	62
Run-off (%)	82	18
Ground water $(1,000\,m^3/km^2)$	165	39

Source: Qu Geping and Li Jinchang, 1992.

cause more pollution. Moreover, coal accounts for more than 70 per cent of China's commercial energy, and the direct use of untreated coal also is another main contributor to heavy pollution.

China has suffered ecological deterioration over a long period of time. Over-farming and grazing have caused the enlargement of desert areas for thousands years in the north, northwest and northeast of the country. Large areas of grassland have been turned into deserts. Among mountainous areas of northern China, forests have been destroyed to meet the demands of construction work and fuel in the cities nearby. Up to 1949 the ecological debts accumulated over time included about 100 million hectares of degraded grassland, 60 thousand square kilometres of human-made desert, and one million square kilometres of land affected by erosion.

Population and environmental problems in recent decades have been strongly affected by these historical and natural factors mentioned above.

Population, Land and Ecological Deterioration

Since 1949 China has achieved some improvement in basic ecological conditions. It is of particular importance that several great watercourses, such as the Yangtze, the Yellow River, the Huai River, and the Hai River, have been brought under control. These rivers used to be the major cause of many natural disasters in China's history. Large-scale farmland improvement has been carried out every year in all of parts of the country, including irrigation and water conservancy construction work, efforts to combat alkalinity and to prevent desertification, and afforestation. The progress of these efforts has greatly enlarged the carrying capacity of the territory.

However, because of rapid population growth, policy mistakes, and other factors, new ecological debts have accumulated continuously. The Great Leap Forward of 1958 was not only an economic disaster, but also caused serious and large-scale ecological degradation. The over-emphasis on grain production in the Cultural Revolution period is another such event.

Consequently, desert areas have expanded by 2,460 square kilometres every year for the last 2 decades, and at present about 27.3 per cent of China's territory is covered by desert (UNDP, 1998).

Farmland is at the core of the relationship between population and natural resources in a country like China. China's arable land accounts for roughly 10 per cent of its total land territory. Of this, only 65 per cent is located on the plain, the remainder being in hill and mountain areas. 63.5 per cent of these arable lands are suffering from various kinds of deterioration in productivity (CAAS, 1995). Total farmland amounted to about 97.3 million hectares at the beginning of the Republic, and was 95.3 millions hectares in 1995. Although the total area under cultivation had remained steady over the years, however, farmland per capita has fallen from 0.18 hectares per capita to only 0.08 hectares.

The steady figure for total farmland is mainly due to the large-scale reclamation of so-called wasteland, which was mainly located in the three river plains of northeast China and the semi-drought areas in north and northwest China. In the south of China, the new farmland was recovered from estuaries, forests, lakes and beaches. Considering that China is extremely short of farmland, the need for large-scale reclamation is justifiable. However, it is also an important factor leading to serious ecological degradation. In Xinjiang there are 1.02 million hectares of new farmland which were reclaimed in the 1950s and 60s, of which about 0.48 million has subsequently suffered from desertification. In the south of China, Hainan and Yunnan provinces have lost 80 and 50 per cent of their rainforest, respectively. In the 1960–70s, large areas of lake in south of China were transformed into farmland (see table 18.2). These practices seriously damaged biodiversity, and cut down the capacity for regulating the flow of water.

China now has only 14.7 million hectares of reserve arable land, most of which is grassland in drought or semi-drought areas where the ecological balance is extremely fragile. In the south of China, the reserve arable land is completely exhausted. In the southeast region, rapid economic growth has caused increasing demands for non-farming land use. The lost farmlands in this region are usually much more productive than those in the northwest, suggesting an even greater loss than is statistically evident.

Table 18.2 Change in number and area of lakes in China

	Beginning of 1950s	*Beginning of 1980s*	*Change*	*% change*
Number of lakes	2,800	2,350	−450	16.07
Total lake area in km²	80,600	71,695	−8,905	11.05

Source: Qu Geping and Li Jinchang, 1992.

Heavy population pressure and the shortage of farmland means that China has to meet demands for agricultural products by raising its output per unit of land and its multiple crops index. The grain output per hectare increased from 1,035 kg in 1949 up to 4,239 kg in 1995. However, the continuous cultivation of farmland has exhausted soil fertility, making agriculture more and more dependent on inputs of chemical fertilizers. On average, Chinese farmers used 378 kg of chemical fertilizers per hectare in 1995. That was 3.5 times the figure for the USA.

According to the statistics for 1994, China had 1.337 million square kilometres of forest, covering 13.92 per cent of its territory. The coverage in 1949 was reported to be just 7.5 per cent – that is believed to be an incomplete figure. However, while the forest areas have been increasing, the total biomass of the forest has continued to decrease, due mainly to the fact that man-made forest has been increasing and natural forest has been decreasing, young growth areas have been expanding and the adult forest shrinking. China's per-capita forest stock volume, averaging 8.6 cubic meters, however, is only 12 per cent of the world average and among the lowest in the world.

There is a shortage of commercial energy in China's rural areas and crop-straw is insufficient to meet the daily fuel demands. This is a very significant cause of ecological degradation because, in order to obtain fuel, trees are cut down and even dry leaves and grassroots are collected. This aggravates soil erosion, and slows down the growth of forest.

China has about 400 millions hectares of grassland. Of this land, about 290 million hectares can be used as pastoral areas. About 160 millions hectares are drought grasslands, semi-desert grasslands and high and cold grasslands. Only 64 million hectares belong to grassy marshlands with higher natural productivity. In the south of China, the grasslands are scattered and it is thus difficult to achieve efficiencies of scale in supporting livestock.

Since 1949 population growth in the grassland areas has been much faster than in the rest of the country. On the one hand, populations in the north and northwest China are Chinese minorities with higher natural growth rates than the rest of China. On the other hand, in the several migration waves since 1949, the grassland areas have all been major immigration zones. For example, Inner Mongolia had a population of 5.67 million in 1947, which had increased to 21.45 million in 1990. According to some estimates, the immigrant population is more than 5 million. Farmers make up the majority of the immigrating population and thus more than 27 millions hectares of grasslands were transformed into farmland. One third of this reclaimed land has become desertified. Meanwhile, the remaining grasslands have been over-grazed and degraded because of growing population pressure (see table 18.3). At present, the total productivity of the 220 million hectares of grassland in North China is only equal to that of 11 million hectares of farmland.

Table 18.3 Human-caused desertification

Human factors	Contribution to desertification (%)
All factors	94.5
Reclamation	25.4
Over-grazing	28.3
Deforestation	31.8
Land use for industry	0.7
Unreasonable water resources use	8.7

Source: Qu Geping and Li Jinchang, 1992.

Environmental Pollution and the Water Crisis

In the first few years of the People's Republic, the industrialization process was in its early stages and there was little obvious pollution. It was the Great Leap Forward in 1958 that made the problem of environmental pollution start to become visible. In 1958, about 600,000 small blast furnaces, 59,000 small coalpits and 9,000 small cement plants were set up utilizing very backward technology, non-skilled workers and traditional management. They were located without planning, so pollution became widespread within a short time. During the Cultural Revolution, the need for management systems in enterprises was denied, and a large number of small and low-grade enterprises were established. The policy emphasis on 'turning the consumer-cities into productive ones' and 'production first' caused dramatic environmental deterioration in urban areas. Many of the enterprises set up during this period have continued to pollute up to the present day.

Pollution of China's atmosphere remains smoke-dominated, with sulphur dioxide and soot as the main pollutants. This mainly results from the fact that more than 75 per cent of all commercial energy in China comes from coal, while oil and natural gas only account for 20 per cent. China consumes about 1.2 billion tons of coal each year, 35 per cent of which are used for power generating and the remaining 60 per cent for other industries and domestically. In 1994 China's energy consumption accounted for 9.6 per cent of the world total, but its CO_2 emissions accounted for 14 per cent. This pattern is expected to be maintained in the near future, even though many efforts have been made to improve the efficiency of energy use.

Due mainly to the rapid process of urbanization and a high concentration of population and industries, the quality of the urban environment is not very encouraging (Li Wen, 1999). Most city river sections are contaminated. Air pollution in a few Chinese cities, including Beijing the capital, has reached an unacceptable level (see table 18.4). Noise pollution in many

Table 18.4 Ambient concentration of suspended particles and SO_2 in $\mu g/m^3$ in major cities

	TSP (daily mean)	SO_2 (daily mean)
Beijing	362.7	88.6
Shengyang	356.9	131.5
Xi'an	444.9	50.0
Shanghai	225.2	63.3
Guangzhou	200.0	45.5
Hong Kong	82.4	n.a.
New York	61.6	37.5
Tokyo	49.9	70.0
WHO recommended limit	60–90	40–60

Source: World Resources Institute, 1996.

cities remains a problem. Solid waste and so-called white pollution have become prominent in the recent years.

The problem of acid rain is serious, with 30 per cent of China's total land area being affected in varying degrees. Industrial solid-waste pollution is another main factor affecting environmental quality. In 1998 the country produced a total of 800 million tons of industrial solid waste, including 9.74 million tons of dangerous waste.

But most serious of all are water pollution and the water resources crisis. In this regard, China faces three major problems: water body pollution, a shortage of water resources, and flooding and damage caused by waterlogging. According to 1994 statistics, in the 110 most important sections of China's seven largest river systems, 39 per cent of the water had medium or serious pollution, with the Huai, the Liao and the Songhua Rivers all seriously polluted. About 80 per cent of wastewater is discharged into rivers without any treatment; about 50 per cent of the water in river sections near cities is so polluted that it is undrinkable. In the north of China, 50–70 per cent of water in some main rivers is completely unusable.

The relationship between water pollution and water shortage is apparent in the north of China. The farmlands in the plain areas along the Huai, the Yellow and the Hai Rivers account for 42 per cent of the total farmland in China but only have 6 per cent of its fresh water resources. In the mid-1980s, the run-off at the estuary of the Hai River was only 5 per cent of the run-off at the beginning of 1950s. In fact, Tianjin City had so little in the way of water resources that water had to be taken from the Luan River, hundreds of kilometres away from the city. It is reported that 300 Chinese cities face the problem of water shortage. Now 50 per cent of the water resources of the Yellow, Huai and Liao Rivers have been exploited. Usually, when river water resources are exploited up to 40 per cent, the ability of the body of water to purify itself begins to decrease and the water environment begins to degrade more quickly. This means that the large watercourses in North China are in danger of extremely serious degradation, as was the case with the Huai River.

In the summer of 1994 a pollution disaster occurred in the Huai River basin. About 200 million cubic metres of foul-smelling black water ran downstream destroying the aquaculture farms of 30 townships, stopping the power stations and waterworks of some cities, and leaving hundreds of thousands of people without drinking water. Although the Huai River basin is not a developed region of China, this was the most serious pollution accident in China's history. The pollution came from a combination of heavy water-resource exploitation and emissions. In recent years a large number of small enterprises have set up in every province in the basin, including Henan, Anhui, Shangdong and Jiangsu, especially small pulp mills, and these small plants are the main sources of pollution in the basin.

The Main Causes of Environmental Deterioration

China has been experiencing a general and serious environmental degradation. Population pressure, natural conditions and fast economic growth can partly explain the problem, but the most significant factor is that the Chinese economy has been growing with low efficiency. For every one unit of GDP created, China's energy consumption is 2.19 times that of the USA, 3.07 times that of the UK, 5.0 times that of Germany, and 5.9 times that of Japan. In China, producing 1 ton of paper uses 450 tons of water, while it takes just 50–200 tons in developed countries. Low efficiency and high wastage result in heavy pollution. For example, the direct use of coal has been the main factor causing air pollution. As table 18.5 shows, in China's primary energy structure the proportion of coal is not only very high, but also continuously increasing. Meanwhile the rate of washed coal has decreased slowly. This results in increased SO_2 pollution.

The main reason for the high consumption of coal is the inefficiency of industry. Rapid economic growth and low efficiency combine to place great pressure on China's energy system. Table 18.6 compares China with other countries in terms of energy intensity of production. It can be seen that the energy intensity of most products in China is around twice the energy intensity in OECD countries. This means that if China can reduce the gap in energy efficiency, significant environmental improvements could be made without direct investment in pollution control.

The sustainable development problems in rural areas are very complex. In rural areas, ecological degradation is usually connected with poverty. According to 1995 statistics, China has about 80 million peoples still living in poverty. It is a small proportion compared with the whole population, but the poor mainly live in areas with serious ecological deterioration. Under the pressures of population growth, these people may be trapped

Table 18.5 Primary energy sources in China

	Coal	Crude oil	Natural gas	Hydro power
	Primary energy consumption (%)			
1978	70.7	22.7	3.2	3.4
1979	71.3	21.8	3.3	3.6
1980	72.2	20.7	3.1	4.0
1981	72.7	20.0	2.8	4.5
1982	73.7	18.9	2.6	4.9
1983	74.2	18.1	2.4	5.3
1984	75.3	17.4	2.4	4.9
1985	75.8	17.1	2.2	4.9
1986	75.8	17.2	2.3	4.7
1987	76.2	17.0	2.1	4.7
1988	76.2	17.0	2.1	4.7
1989	76.0	17.1	2.0	4.9
1990	76.2	16.6	2.1	5.1
1991	76.1	17.1	2.0	4.8
1992	75.7	17.5	1.9	4.9
1993	74.7	18.2	1.9	5.2
1994	75.0	17.4	1.9	5.7
1995	75.0	17.3	1.8	5.9

Source: *The Statistics Yearbook of China* 1992, 1994, 1996.

Table 18.6 Comparison of energy intensity

Products		Energy intensity in China (A)	Energy intensity in other countries (B)	A/B
Steel and iron	kg SCE/ton	1,034	629 Japan	1.64
Coke	kg/ton	577	466 Japan	1.29
Aluminium oxide	kg SCE/ton	1,916	414 Germany	4.62
Iil industry	kg SCE/ton	148.9	71.68 Japan	2.08
Synthetic ammonia	kg SCE/ton	1,290	930 USA	1.38
Ethylene	kg SCE/ton	1,276.1	670 USA	1.90
Cement	kg SCE/ton	201	113.2 Japan	1.78

Source: *Energy Development Report in China*, 1994.

in a cycle of poverty and ecological degradation unless they receive outside help.

After rapid economic growth in the last two decades, China's economy is now developing on a massive scale. In the future there is a pressing need for environmental policy to improve industrial energy and resource efficiency, reduce waste and to protect the quality of the environment.

Efforts at Environment Protection

The Chinese government has responded actively to the United Nations Conference on Environment and Development (UNCED) by taking the lead in formulating the first national Agenda 21 in the world. This document is titled China's Agenda 21 – a White Paper on China's Population, Environment and Development in the 21st Century, drafted in 1994. This document integrates the principles of UNCED with China's particular conditions and needs. Soon afterwards, many local governments also adopted plans for implementing China's Agenda 21 in their respective jurisdictions. In July 1996 the State Council held the Fourth National Conference on Environmental Protection in order to implement the sustainable development strategy. Subsequently, the 'Decision of the State Council on Issues Related to Environmental Protection' was drafted, so was a specific sectoral plan 'The Ninth Five-Year Plan and the Long-term Objectives for the Year 2010 on Land and Water Conservation'.

Environmental protection is viewed as a basic national policy; environmental measures are to be integrated with a general economic development strategy, macro policy arrangement and action incentive structure. This view has been reflected in the continuous adjustment and transformation of environmental policies. So far China has developed a fairly complete system of environmental policy, covering environmental pollution control and environmental conservation. While some of these policies are prescriptive and mandatory in nature, others are more in the way of guidelines (Xia, 1999). In addition, by 1997, four environmental laws, eight resource management laws, more than 20 administrative regulations on the environment and resource management, and more than 260 environmental standards had been promulgated to form the basis of the legal system for environmental and resource protection.

Following the general guidance of the above-mentioned acts, China formulated its 'Programme for Controlling the Total Amount of Major Pollutants during the Ninth Five-Year Plan' and 'Trans-Century Green Project'. While the former advocates strict regulations in controlling the total amount of 12 major pollutants, the latter is specifically targeted at areas and river basins with critical pollution problems and fundamental environmental problems. The project puts its primary focus on water pollution and acid rain in southwestern, central, southern, and eastern China, as well as air pollution in 30 key cities.

China's environmental policy and programmes have made great achievements and been basically successful, but, nevertheless, certain structural defects remain and quite large difficulties still have to be faced.

First, the up-trend of environmental pollution has been slowed down somewhat over the last 10 years or more, a period in which China's economy grew by an average of 9 to 10 per cent. In other words, China is on the right track to avoid the common trap of rapid economic growth accompanied by rapid deterioration in environmental quality. Industrial output in 1995 was more than three times the output in 1981, but wastewater emissions from urban industry have been kept at a level of 25 billion tons per year. The same is also true for the SO_2 level in cities. Changes in China's economic structure, better access to the global capital and technology markets, enhancing the energy efficiency of the Chinese economy, strengthened economic ability to cope with environmental problems, and so on, are the main forces that contributed to this trend.

Second, investment in pollution control has increased year by year from 2.5 billion RMB yuan in 1981 to 40.8 billion in 1996, and will, reportedly, reach 65 billion by the year 2000. Likewise, the share of environment investment in total GDP will have increased from about 0.5–0.8 per cent to 1.5 per cent by the end of the twentieth century (IMCAES, 1998). China will also invest 500 million yuan annually to develop environmental protection technology, equal to the total amount spent in the Eighth Five-Year Plan period (1991–5). Increased investment enables the government and society to conduct large-scale environment protection projects. For example, by the end of 1998, China had completed the comprehensive treatment of 78 million hectares of land affected by soil erosion, and built more than 100 million water storage and soil conservation projects. Water and soil conservation projects in the seven major river basins now cover more than 800 counties in 26 provinces and more than 20,000 small watersheds, with projects on more than 5,000 small watersheds completed (*Beijing Review*, 1999).

Third, the pollution abatements in some priority areas and partial regions have made obvious progress. In the Huai River basin more than 5,700 enterprises have been closed, thus decreasing the pollution burden on the basin by 40 per cent. The Chinese Government has introduced a series of measures to intensify the prevention and control of acid rain pollution. As a result, pollution in the central China acid rain zone was minimized, although the overall pollution pattern in other parts of China remained unchanged. In addition, the environmental quality in some cities has been to a certain extent improved.

Fourth, the basic systemic structure for environmental protection, including management, monitoring, scientific research, education, etc. has been established and started to function effectively (Xia, 1999) Environmental protection agencies (EPA) have been significantly strengthened. By the end of 1997, there were 9,207 environment protection agencies (units) across China staffed by 103,180 personnel. A comprehensive legal system, consisting of laws, regulations and various other policy measures, has also been developed. It could be said that China's environmental situation would be much more serious without this intervention.

However, China's environmental policy and programmes are under increasing challenge. Since 19980s many official assessment about China's environmental situation have been conducted. The usual conclusions of these evaluations are 'partially improved while getting worse overall' (1988), and 'there is still no change in the conditions of environmental pollution and ecological wreck in quite a large number of regions, and in some of them the situation has even got worse' (1998). It seems most likely that the effectiveness of environmental policies has always been offset by newly produced environmental pressures, and the environmental policies have simply preserved the environmental status from sharp decline, that is, they have not had the ability to radically change the serious state of the environment (Xia, 1999).

China continues to face a grim environmental situation. Air pollution will continue to worsen for the next 20 to 30 years, hitting its high point 2030, as coal consumption will increase by 60 to 85 per cent by 2020. More areas will be affected by acid rain. Water pollution will not be brought under control in the short term, as increasing amounts of domestic sewage more than make up for waning industrial wastewater discharges. The average person produces 440 kilograms of garbage annually, a figure rising by 8 to 10 per cent every year. Accumulated garbage covers about 500 million square meters of land across the country, and citizens in more than 300 cities are threatened by its polluting effects (Li Song, 1999) With sustained population growth and an improving standard of living, four times as much garbage will be produced in 2020 as in 1996. Furthermore, in the next 20 years China's cities will face great environmental challenges as half of the total population moves to urban areas.

There are a few major shortcomings in China's environment protection. First is the separation of the economic development decision-making process from the requirements of environmental protection – or the very loose integration of the two. The issue of who is responsible for environment protection has been a continuing subject of debate during the last few decades (Sinkule, 1995). Second is the lack of sufficient resources to make effective supervision and restrictions against environmentally harmful actions (Xia, 1993). Furthermore, the low awareness of the public on environment issues also makes protection of the environment a government function. Mass participation is only a recent phenomenon and needs to be promoted sincerely in the near future.

References

Administrative Centre for China's Agenda 21 (1997), *The National Report on Sustainable Development*, Beijing.

China Academy of Agricultural Sciences (1995), *China's Arable Land.* Beijing: China Agricultural Sciences Press.

China Environment Yearbook, 1994–1997. Beijing: China Environment Yearbook Publishing House.

Environmental and Resources Protection Committee of the National People's Congress (1997), *Collection of Laws and Regulations for Environment and Resources.* Beijing: China Legislation Press.

Institute for Management of China Academy of Environmental Sciences (IMCAES) (1998), *An Economic Study of China's Industrial Pollution Control,* (Research Report), Beijing.

Li, Song (1999), China's environment in the 21st century, *China Daily,* No.1.

Li, Wen (1999), China's environment conditions in 1998, *Beijing Review,* 42 (28).

NEPA, State Planing Committee, State Economic and Trade Committee (1996), *Transcentury Green Projects Programme (Phase I, 1996–2000).* Beijing: China Environmental Sciences Press.

Qu, Geping and Li, Jinchang (1992), *China's Population and Environment.* Beijing: China Environmental Sciences Press.

Sinkule, B. and Ortolano, L. (1995), *Implementing Environment Policy in China.* Westport, USA: Praeger.

State Development Planning Committee (1997), *1997 Report on Population, Resources and Environment in China.* Beijing: China Environmental Sciences Press.

State Environmental Protection Administration (SEPA) (1998a), *Transcentury Environmental Protection in China.* Environmental Working Communications, no.8.

State Environmental Protection Administration (SEPA) (1998b), *Report on Environmental State in China 1997.*

UNDP (1998), *China Human Development Report.* Beijing.

Xia, Guang (1993), On the market-oriented agent system for environmental right and benefit. In *Towards the 21st Century.* Beijing: China Environmental Sciences Press.

Xia, Guang (1999), *Structure and Priorities of China's Environmental Policies.* Paper presented at the 5th Hiroshima-Shanghai Conference on Sustainable Development, 4 October 1999, Hiroshima University, Japan.

19   The Population of China: Prospects and Challenges

Zhai Zhenwu

On 15 February 1995, China officially announced to the world that the day of a population of 1.2 billion in China had arrived. This meant that despite a variety of measures and efforts by the Chinese Government, China had breached, five years in advance, the demographic limit set by the Chinese Government in the early 1980s for the year 2000. At the same time, the world population reached 5.7 billion, and China accounted for 21 per cent of the world total, implying that changes in the fertility, mortality, and growth rate of China's population would have a considerable impact on the world population. For example, as far as the demographic indicators for developing regions are concerned, levels differ greatly depending on whether the category includes China or excludes it. In 1995 the developing regions including China had a crude birth rate of 28 per 1,000, a natural growth rate of 1.9 per cent, a total fertility rate of 3.5, and an infant mortality rate of 67 per 1,000. This contrasted with figures of 31 per 1,000, 2.2 per cent, 4.0, and 72 per 1,000, respectively, for the developing regions excluding China (Population Reference Bureau Inc., 1995).

Over the decades, China has had moderately high population growth rates as compared with other developing countries. The first post-war baby boom occurred in China in 1952–7 with an average natural growth rate slightly over 2 per cent, then a baby bust followed in 1958–61 as a result of the rise in mortality and drop in fertility due to the dramatic change in the rural economic system plus 3 years of natural disaster that led to great hardship for the Chinese people. The natural growth rate remained at around 20 per 1,000 in 1962–73, which was the second baby boom, and dropped to below 20 per 1,000 after 1974, reaching a low of 11.61 per 1,000 in 1979. It vacillated between 11 and 14 per 1,000 thereafter. In 1995 a growth rate of 12 per 1,000 was recorded (China Population Information and Research Centre, 1996).

Despite the relatively low natural growth rate, new increments to China's population have been substantial in recent years, about 14 million per annum, increasingly aggravating the population pressure on China's economy, resources, and environment. Though fertility in China has been

reduced to below-replacement level, a large increase of population continues as a result of the huge momentum of population growth. It is a
critical and arduous task for China to deal with a range of demographic
challenges during the period of dramatic population change, before as well
as after zero population growth comes.

Assessment and Assumptions of Future Mortality and Fertility in China

The growth of the population in a given country is exclusively determined
by fertility, mortality, and international migration. As there has been little
international migration in China, population growth in China is largely
attributable to the levels of fertility and mortality.

Estimation of future level of mortality in China

The life expectancy at birth of the Chinese population in 1990 reached 67.5
years for males and 71 years for females. Based on empirical patterns of
mortality transition, improvement in life expectancy becomes increasingly
slower after reaching a certain level. According to United Nations estimates, when life expectancy at birth reaches 67.5–70 years, its quinquennial
gains stand at 0.75 years for males and 1.80 years for females; when it
reaches 70–72.5 years, the quinquennial gains become 0.45 and 1.40, respectively, and will further slow down to 0.20 and 1.00 years when the life
expectancy reaches 72.5–75 years. In this chapter, the male and female life
expectancies of China for the projection period 1995–2060 are calculated
based on these empirical values, taking the 1990 census life expectancy as
the starting point. On the basis of the mortality pattern of the 1990 life table,
a series of life tables are generated by various levels of life expectancy, and
changes in mortality of China are assumed to follow this series.

Estimation of future level of fertility in China

Before 1970 the total fertility rate (TFR) of Chinese women was maintained at over 5.5, but since 1972, when China vigorously carried through
its family-planning programme, the TFR has fallen rapidly, from 5.8 in
1970 to 2.6 in 1981. The government-sponsored family-planning programme
has played a crucial role in bringing down Chinese fertility, however the
fundamental changes that have taken place in China's socio-economic
structure since 1950 have also undermined the previous reproductive
norms and paved the way for the fertility reduction in the 1970s. The trend
of fertility reduction that occurred in the 1970s did not continue into the
1980s. Despite government efforts to implement family planning, the potential for fertility decline created by the socio-economic changes of 1950–70s
tended to be exhausted by the 1980s, leading to a TFR fluctuating between
2.3–2.9.

However, fertility transition in China has been quietly but strongly influenced by the more fundamental changes in China's socio-economy generated by economic reform in the 1980s, and as a result, the early 1990s witnessed another nationwide downward trend in fertility, with the coastal 'opened-up' areas at the fore. According to official statistics, the TFR was reduced from 2.3 in 1990 to as low as 2.0 in 1992 (Yao Xinwu, 1995). Despite the marked reductions in fertility in the early 1990s, it has been a subject of controversy whether fertility in China was dropping as rapidly as indicated by the official statistics. Some demographers argue that the official birth statistics are subject to serious undercounting, and the TFR in the mid-1990s was estimated at 2.1–2.2 (Zeng Yi, 1995).

Three kinds of fertility assumptions are provided in light of the above-described trend of fertility change in China. First, a somewhat loose one, envisaging that TFR remains constant at the 1990 level (2.3) throughout the period. Second, an assumption of an increasingly reduced rate of decline, positing that the TFR approaches the replacement level in 1995 and goes down further to 2.0 by 2000 and after that to 1.8, a value representative of the average level of the more developed countries in the early 1990s. (In 1990, for example, northern Europe had a TFR of 1.8, western Europe 1.7, and North America 2.0.) This assumption is based on the idea that various factors conducive to fertility decline will result from improvements in China's socio-economic structure, level of urbanization, per capita income in rural areas, and primary and secondary education, and on a cautious assessment of the speed of fertility decline. Comparing it with the official birth statistics for 1990–95, we will find this assumed rate of fertility decline is slower than the one indicated official data. The third assumption posits a TFR of 2.3 in 1990 that drops to 1.7 by 2000 and thereafter stays unchanged. It is based on the determination and capacity of the government to carry out its family-planning policy, plus the significant impact on people's reproductive behaviour of the fundamental changes that have occurred in China's socio-economic situation since the opening and reform. It assumes the possibility of a realization of a speed of fertility decline consistent with the official birth statistics for 1990–5.

Hence, projections of China's population have been prepared for 1995–2060 on the basis of the 1990 census age structure of population under one set of mortality assumptions and three scenarios of fertility assumptions.

Future Growth of Population and Changes in Age Structure in China

Figure 19.1 displays the trends of population growth in China under three variant projections. The high-, medium-, and low-variant projections will, respectively, lead to a population of 1.311, 1.293 and 1.276 billion by the end of this century, and this population difference becomes increasingly large as time goes on.

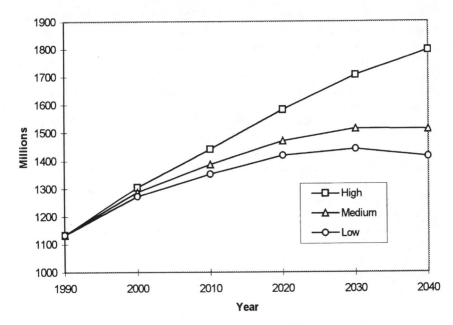

Figure 19.1   Projected total population of China, 1990–2040

According to the high-variant projection, TFRs are kept constant at 2.3, which is still considerably lower than the average level of 3.1 for the world, and similar to the level of Thailand and Iceland in 1995. The growth of the total population will never cease, despite a steady decline in the growth rate – with the decennial rate of population growth going down to 10.1 per cent, 9.4 per cent, 7.5 per cent and 5.2 per cent, respectively after the turn of the century and thereafter reducing further. The total population will ultimately tend towards stability with a low rate of growth. This is an alarming scenario in which population of China will be growing slowly but endlessly.

Under the medium- and low-variant projections, as a result of the momentum growth of population, the total population in China will not stop growing until the 2030s when a historical zero growth finally comes and the total population reaches a peak of 1.519 or 1.441 billion respectively. After that China will experience negative population growth, leading to a gradual decline in the total population. The fertility difference between the medium- and low-variant projections is only at 0.1–0.3, but the maximum population will differ by 78 million. Therefore, policy-makers should make it their first priority to endeavour to create a good socio-economic environment in order to promote the transition of fertility norms among the people, particularly the peasants, and to achieve zero and negative population growth as indicated by the low-variant projection.

On the basis of the actual changes in Chinese fertility since the beginning of the 1990s, most experts believe that the population of China will most probably move in a direction between the medium- and low-variant projections. Table 19.1 shows the projected total population on the medium- and low-fertility assumptions, and table 19.2 presents the demographic indicators of natural growth deriving from the medium-variant projection.

The population growth rate will put China into the category of low-growth countries after 1995: the growth rate of 1995–2000 will be similar to the 1995 rate in Iceland, the United States, New Zealand and Australia, and after 2005, 2015 and 2025, it will be similar to the 1995 rates of Canada, Japan and Germany, respectively. By that time, China will have completed the demographic transition and will start its unprecedented negative population growth, which means that population pressure will ease up in China, a country that has long been perplexed by a excessively high density of population.

If affected by low fertility (the medium-fertility assumption) during the coming half century, China will undergo a considerable change in the age

Table 19.1 Projected total population (millions)

Year	Medium	Low
1995	1,217	1,212
2000	1,288	1,272
2010	1,387	1,352
2020	1,470	1,419
2030	1,515	1,441
2040	1,512	1,415

Table 19.2 Birth, death and natural growth rates (medium projection)

Year	Birth rate	Death rate	Natural growth rate	Doubling time
1990–95	20.9	6.8	1.4	50
1995–2000	18.1	6.7	1.1	63
2000–05	15.1	6.8	0.8	87
2005–10	13.7	7.1	0.7	99
2010–15	13.5	7.4	0.6	116
2015–20	12.5	8.5	0.4	173
2025–30	10.8	10.4	0.0	—
2035–40	10.7	11.6	–0.1	—

Table 19.3 Change of population age structure

Year	0–14 (%)	16–64 (%)	65 + (%)	Old-child ratio (%)	Median age
1990	27.6	66.8	5.6	20.3	25.3
1995	27.4	66.5	6.1	22.3	27.3
2000	26.6	66.6	6.3	25.6	29.4
2005	24.1	68.6	7.3	30.3	31.6
2010	21.1	70.9	7.8	36.4	33.5
2020	19.0	70.2	10.9	57.4	36.1
2030	17.9	67.6	14.5	81.0	39.4
2040	16.3	64.3	19.5	119.6	42.2

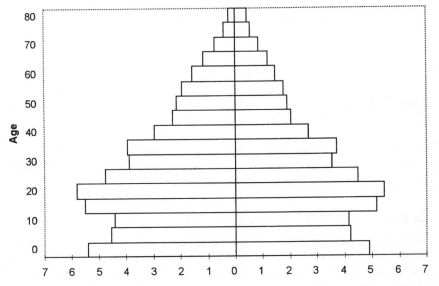

Figure 19.2 Population pyramid of China, 1990

structure of its population, characterized by a continuous decline of the child proportion and a steady increase of the proportion of old people. As shown in Table 19.3, the child proportion will go down from 27.6 per cent in 1990 to 16.3 per cent in 2040, while the old proportion will rise from 5.6 per cent in 1990 to 19.5 per cent in 2040. The population structure will thus become an aged type (figure 19.2 and figure 19.3). By general demographic criteria, populations that have a median age over 30 years, an old-child ratio over 30 per cent or an old (65–) proportion over 7 per cent are classified

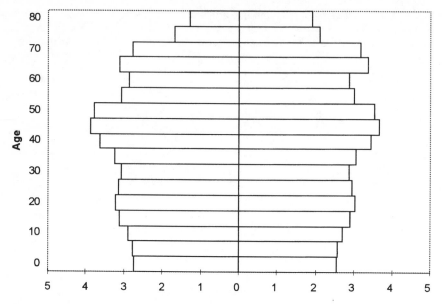

Figure 19.3 Population pyramid of China, 2040

as old-type populations. Table 19.3 clearly demonstrates that China will acquire an old-type population by 2005, when these three indicators reach 31.6 years, 30.3 per cent and 7.3 per cent, respectively.

Challenges in the Field of Population

Since the 1980s, the economy of China has undergone a fundamental change characterized by a high rate of economic growth. Experts predict that, if both the domestic and international environment remain stable, the Chinese economy will, on the one hand, continue growing at high speed with an annual rate of growth in GNP of 8–10 per cent, and, on the other, undergo an all-round restructuring, transforming it from a growth-from-expansion into a growth-from-benefits economy. The early decades of the twenty-first century will be crucial for China if it is to realize its economic take-off completely. But China will then still face great challenges in the population field in addition to those from the process of economic take-off itself. The most important of the population challenges are the following.

The total population of China will reach a historical peak of 1.519 billion

The population–resources relationship in China has all along been strained. The food problem in particular has received world-wide attention, the basic question being: 'Can China feed itself?'. The total arable land in China in 1990 was less than 100 million hectare, and the per capita arable land was only 1.3 mu. Due to population growth, economic development and the rising of level of urbanization, the arable land in China is being reduced annually by 400–500 thousand hectare. Even supposing that newly culti-vated land in each year is enough to compensate for this reduction, the per-capita arable land in China will still decline to 1.22 mu in 1995, 1.15 mu in 2000, to roughly 1 mu in 2020. The intensity with which land is cultivated increases with the decline of the land area, thus the capacity for holding the ecological balance in land tends to weaken as the population pressure becomes increasingly large, given that the production-sustaining capacity of land is limited at a certain technological level. The population–food–land equation thus remains to be resolved.

Mere subsistence is not the goal of development in China. The Chinese government has been endeavouring to develop the economy, carry forward industrialization, rapidly improve the living standards of the people, and has been striving to become moderately developed by the mid-twenty-first century. A necessary condition for achieving this development goal is an ample supply of various energy and natural resources, but with the huge population of China, every small rise in living standards may be obtained at the cost of large increases in the consump-tion of these resources. It is undoubtedly a great challenge to economic take-off in China that the country wants to complete the industrialization process at the time as reaching its historical maximum population – 1.519 billion.

The labour supply will reach a maximum of 893 million

The huge and growing labour force is an important component of economic development, but at the same time puts a tremendous pressure on employ-ment. As projection results show (table 19.4 and figure 19.4), the labour force population, comprising males aged 16–59 and females aged 16–54, will

Table 19.4 Projected labour force (millions)

Year	Male (16–59)	Female (16–54)	Total	%
1990	364.03	317.13	681.16	100
2000	410.97	363.26	774.23	114
2010	465.89	399.55	865.44	127
2020	483.55	409.50	893.05	131
2030	460.34	379.71	840.05	123
2040	446.97	375.99	822.96	121

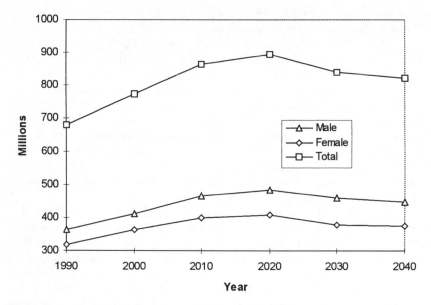

Figure 19.4 Projected labour force of China, 1990–2040

average 820 million in the coming 50 years, reaching a maximum of 893 million in 2020 and slowly declining again to 820 million in 2040. The labour force may be partly absorbed by secondary and higher education institutions, but what remains is still an enormous contingent that exceeds the total labour force in all the developed countries combined. Moreover, over a half of these 800 million labourers will be transferred from the countryside to the cities and towns, thus intensifying the employment pressure there.

China is predominantly an agricultural country in the sense that 80 per cent of the Chinese population live in rural areas. Proceeding from an agricultural country towards an industrialized one, there appears to be a general law that the agricultural population will decline with rural labour being transferred to the cities and towns where industrialization is centred. However, it is very difficult for the rural labour force to be absorbed by the cities and towns even when there are no increments to the labour force as a whole (if, for example, it were maintained at the 1990 level of 681 million). Furthermore the total labour force is experiencing a steady increase towards its projected historical maximum value, which is, no doubt, another significant challenge that China faces in dealing with the employment problem while improving labour productivity in the process of realizing industrialization in the twenty-first century.

The population of China is ageing rapidly

An increasingly prominent problem, in the context of the growing population and labour force, is one relating to the age structure, that is the rapid ageing of the population. As a result of the population control policy of recent decades, the pressure of total population has been eased, but the shape of the population pyramid has changed from one with a wide base and a narrow top to one with a narrower base and a bulge towards the top (see population pyramid in 2040 in figure 19.3).

By 2018, the proportion of the old population aged 65 and over will break 10 per cent. Thereafter the old proportion will grow at an annual rate of 2 per cent, and, by 2040, the old proportion will reach 19.5% with a total old population of 300 million. When the old are defined as those aged over 60, the old proportion will reach as high as 25 per cent with a total old population of 380 million, that is to say, there will be 1 old person (aged 60 and over) in every 4 people in China.

Currently Sweden has the oldest population in the world, the old proportion in 1990 was about 18 per cent. It took almost 200 years for Sweden to go from a young to an old-age structure of population, while demographic transition in China has been a considerably accelerated process characterized by completion of the age structure transition within 40–50 years due to the dramatic fall in fertility. The major features of population ageing in China are the huge old population coupled with the rapid ageing of population. However, the current old-age support system in China also raises issues of concern. The first of these is the low coverage of the country's pension system, with the rural elderly unable to obtain any pensions. The second is the fact that the current medical care system in urban areas is unable to meet the demands for medical care of the urban elderly, while the medical expenses of the rural elderly are mostly borne by their families and children since rural areas are not covered by the state medical care system. A huge amount of funds need to be put into improving these systems, although funds are the biggest shortage for the developing economy of China. Therefore, one of the most critical problems that China faces in the twenty-first century is to confront successfully the ageing issue by formulating and perfecting various old-age support systems while developing the economy speedily. Both need considerable levels of investment.

China is a country full of promise for the twenty-first century, during which it will realize economic take-off. But population ageing in the twenty-first century will raise a difficult problem. By the middle of the next century, the total population of China will still be the largest in the world, but the age structure of population will make it one of the countries with the oldest populations. And all of this will happen before the completion of the modernization process. China will have to confront the pressure from a maximum total population plus a maximum total labour force, at the same time as facing the challenge of population ageing.

References

China Population Information Centre (1996), *China Population Data Sheet 1996.* Beijing.

Population Reference Bureau Inc. (1996), *World Population Data Sheet 1996.*

Yao, Xinwu (1995), *Fertility Data of China.* Beijing: China Population Press.

Zeng, Yi (1995), Has China's fertility in 1991 and 1992 been far below the replacement level? *Population Research*, May.

20 Hong Kong Special Administrative Region: Waves of Chinese Immigrants and their Children

Lui Ping-keung

Introduction

This chapter discusses the demographic aspects of one small geographical region of China, namely the Hong Kong Special Administrative Region. A sharp focus on this small region allows discussion of the close connections between demographic dimensions and the economic, social and cultural configurations to which they are so closely linked.

The majority of the Hong Kong population is of mainland Chinese origin. The story of Hong Kong thus rightly begins with the history of immigration from mainland China after World War II. Successive waves of Chinese immigrants encountered difficulties as well as opportunities for economic and social development – so that one can argue that the Hong Kong of today is a product of the blood and sweat mainly of Chinese immigrants and their children supplemented by the goodwill and shrewdness of the British colonists.

Immigrants usually integrate into the receiving region or country at the lower end of society – as the ruled – and consequently the changing attitudes and policies of the government of the receiving region or country – as the ruler – are highly influential to the subsequent course of integration. Hong Kong is not an exception to this rule. Not until the late 1980s was some form of democracy introduced in Hong Kong. The administration was until the last day of the British rule 'executive led', that is, it had the largest share of power. The official accounts of immigration and related matters therefore form the background for a proper understanding of the local demographic, economic, social and cultural intricacies.

The study of the attitudes to immigration are greatly aided by the annual reports published by the Hong Kong government. Standpoint aside, these reports are to a high degree factually accurate. Because of the complexity of the subject, this chapter will be organized in a fairly unusual manner.

Extracts from the annual reports will be reproduced below to form a chronicle of demographic events and figures. Contemporary remarks, observations, opinions and thoughts on demographic as well as economic, social and cultural matters will be inserted where relevant. The official chronicle and our own narrative will therefore intermingle.

Since the government also describes economic, social and cultural developments in Hong Kong and expresses its position and views on such matters in its annual reports, our own narrative very often involves quotations of these position and views. Also, there is a moderate research literature accumulated over the past three decades but very little belongs to demography in the proper sense. Much of this literature is given in the notes at the end of this chapter so that the text can remain more readable.

1931–1950: The First Surge of Chinese Immigrants

1931 'The last census [before World War II] in 1931 found the total [civilian] population to be 849,751.' (Hong Kong Government, hereafter HKG, 1962).
1941 'An unofficial count by air-raid wardens in 1941 before the Japanese invasion of the [then] Colony put the population at about 1,600,000.' (HKG, 1962).
1945 'This number was greatly reduced during the [Japanese] occupation and it is estimated that the total amounted to less than 600,000 when the Colony was liberated in August 1945.' (HKG, 1962).
1946 '[But] by the end of 1946 the population [was believed to have] again risen to 1,600,000 as people in search of safety and work drifted back to the Colony from South China.' (HKG, 1958).
1949 'In 1949, when the new regime [that is, the People's Republic of China] was securing its grip upon the mainland, immigration became a torrent.' (HKG, 1958) 'An assessment of the population in September 1949 by the then Department of Statistics put the total at 1,857,000.' (HKG, 1962).
1950 'By April 1950 the population had increased to 2,360,000. A survey . . . in 1954 estimated that the rise in population included some 700,000 refugees.' (HKG, 1958) 'By May 1950, . . . [restrictions] on entry from China were inevitable.' (HKG, 1957).

'The problem of a vast immigrant population' and housing

'The first influx fled from the shattered economy and threat of famine which followed the Pacific War. The people who followed in the second influx voted with their feet against the new regime which was established when the Nationalists withdrew to Formosa (Taiwan). In either case the immigrants sought in Hong Kong something sufficiently important to themselves to necessitate the abandonment of their homes, the severance of family ties and the renunciation of traditional allegiances. No one will ever know what it cost them to abandon the land on which their ancestors had made their living. They were not denied what they sought, and Hong Kong accepted the burden which they brought with them in the name of humanity rather than because it had any special standing in the matter other than the accident of contiguity' (HKG, 1957). It should be mentioned that until May 1950, Chinese were allowed to move freely across the border with China. The two influxes were the first surge of Chinese immigrants after World War II; the government simply did not know that there were more influxes to come. The great majority of this first surge did not emigrate elsewhere and finally settled in Hong Kong.

It was 'the problem of a vast immigrant population' as negatively phrased by the Hong Kong government: 'Finance, education, medical and health services, social welfare, prisons, police, industry, commerce, labour relations, land policy, housing, agriculture and fisheries, political relations – even the law itself – all bearing the unmistakable surcharge (in a few cases an almost obliterating surcharge) of this single problem' (HKG, 1957). 'The immigrants were homeless . . . [When] virtually all the vacant urban sites . . . had been over-filled with their flimsy insanitary shelters, they moved into the hills with which the cities [of Hong Kong and Kowloon] are surrounded and hung their shacks in deep festoons over rocks bared by the war-time search for fuel. But always they crowded in on the town, for there alone lay the hope of rice for tomorrow.' Though the language is old-fashioned and the tone melancholy, the government managed to identify one of the key problems that shaped and still is shaping Hong Kong – housing.

The government was forced to act by a number of fires in the squatter areas. 'The story begins in January 1950, when the population was estimated at 2,360,000 and the squatters at something over 300,000. In that month a squatter fire took place in Kowloon City which rendered 20,000 persons homeless' (HKG, 1957). This was followed by more serious squatter fires in the following four years in various different areas. The worst one finally came. 'In the Shek Kip Mei fire of 25th December, 1953, over 50,000 persons lost their homes. This constituted a crisis of the first order . . . Immediate decisions were taken . . . that the Government would itself build and finance the resettlement buildings . . . [This] implied that Government would, from now on, itself enter the field of resettlement using public funds and its own constructional resources' (HKG, 1957). 'In July 1954 . . . the third worst squatter fire in the Colony's history occurred at Tai Hang Tung . . . 24,000 people lost their homes . . .'. 'The conclusion . . . was eventually accepted by Government . . . that . . . resettlement must take place in buildings of six or seven storeys. . . . This decision put the final seal on Government's strange new role of financier, contractor and landlord to a potential 20% of the population . . . at a rent which resettled squatters could afford to pay . . .' (HKG, 1957).

From this urgent beginning, the public rental housing scheme grew phenomenally in the 1960s, 1970s and 1980s, as the population also grew phenomenally through immigration and natural increase. The quality of public rental housing improved throughout the period (see Lai, 1993). Rents remained low in comparison with private housing. The demand for public rental housing was never low. Eventually, all those who were unable to provide adequate housing for themselves, including non-squatters, were eligible for public rental housing. There was a household income upper limit for eligibility for public rental housing but there was no tenancy limit once a family or household was allotted a housing unit, even if in the future its income exceeded the eligibility criterion or the family or household concerned owned a private housing unit. In this sense, it was a life-long entitlement to public rental housing benefits. It enabled a high rate of wealth accumulation at the family or household level among public rental housing tenants. At the same time, private rental housing was never inexpensive. In a colony of Chinese immigrants, most of whom were of limited financial resources at the time of their arrival, public rental housing was a tremendous welfare benefit. In fact, over the decades it created one of the two deepest societal cleavages in Hong Kong (The other being property ownership – explained below). The public rental housing tenants congregated into a vested interest, a political force.

The year 1977 saw the beginning of the Home Ownership Scheme (HOS) in which the government built housing units for sale at reduced prices to eligible families or households. The eligibility criterion was higher than that for public rental housing. The background to this scheme was that since the early 1970s the prices of housing units built by private developers had grown to such a level that many families could not afford to buy one. Throughout the 1980s and into the 1990s until now, the prices of private housing units have continued to rise phenomenally. Accordingly, this scheme continued to grow. Furthermore, the idea was expanded to develop a similar scheme called Private Sector Participation Scheme (PSPS) in which private property developers were invited to join hands with the Housing Authority to build and sell housing units to eligible families under the HOS scheme.

The steeply rising prices of private housing units, the less steeply rising and yet relatively high rent of private housing units, the limited supply of HOS/PSPS housing units and limited supply of land, the limited availability of public rental housing due to a very low level of circulation of tenants, together with a most intense – and to some outsiders enigmatic – desire to own real estate among the local population, pushed the prices of housing units to such a high level that real estate became the major if not the only asset of most better-off families. Property ownership became another deep societal cleavage in Hong Kong.[1] Together with public rental housing tenancy, it basically stratified local society into four classes: namely, private housing tenants who cannot afford to buy a housing unit, public rental housing tenants who cannot afford to buy a housing unit, people who own some property or properties of lesser value, and those who own properties of high value.

There are no quantitative data on how various waves of Chinese immigrants fared in this active social stratification. Many of them benefited and probably still benefit from public rental housing. By and large, it is believed that the earlier waves fared better than the later ones, since the former could ride the tide of the always improving local economy (over a long period of four decades) longer than the latter and hence could enjoy a greater share of the wealth created by the local population. Furthermore, children of the former could also have the opportunity to ride the same economic tide and to benefit from the improving social infrastructure in areas such as education and welfare.[2] The cumulative effect of the timing of immigration at the family level is probably vast and deep.

1951–1961: Sustained Immigration from China, and the First Census after World War II

'[By] the end of 1956 [the population had risen] to an estimated 2,500,000 . . .' and further increased to 2,919,000 at the end of 1959. 'The estimated increase during the course of the year 1959 was 113,000; of which 84,329 represents natural increase. A stable or reasonably stable population is clearly not yet in sight; for, even if immigration could be wholly checked, the annual rate of natural increase would still stand at approximately 3% . . .' 'The population problem is complicated by illegal immigrants who may have added considerably to the estimates given above and who, during 1959, are thought to have numbered very approximately 30,000, though this figure has not been taken into account [that is, in the calculation above.]' (HKG, 1960).

'In the spring of 1961 Government carried out the first census in thirty years of the population of Hong Kong . . .' 'The . . . figure [of the census] was . . . 1,610,650 males and 1,522,481 females, 3,133,131 in all, including 3,483 persons classified as transients' (HKG, 1961). 'It was a young population. . . . 40.8% . . . are under 15 and . . . no less than . . . 16.0% . . . are under five. But from our young people there is half a generation missing. In the age group from 13 to 27 inclusive, that is those born between 7th March 1934 and 6th March 1948, there are about 177,000 males and 164,000 females less than the other proportions of the population require . . . ; and this deficiency is most marked in the ages 15 to 20 inclusive, where the numbers missing exceed the survivors. This is the generation on whom lay most heavily the privations of the years 1938 to 1945, and the scar which marks the amputation of half their number is a silent witness to the ghastly irrelevance of warfare. . . . The privations of the war and the occupation lay heavily on the aged, too . . . No doubt the culling of the feeble and chronically sick from those ages which were neither young nor old had something to do with the average fitness and health of the survivors.'

'It is a healthy population. The death rate is low and getting lower. The birth rate is moderately high and the rate of infantile mortality is remarkably low. These factors point to a steadily rising rate of annual natural increase, and there are several indications that this rate of increase will steepen sharply in four years' time. . . .' 'It is becoming a settled population. In former censuses the sex ratio . . . showed a large excess of males, because large numbers of men came here from China without their families and returned to their villages in 'Canton more far' when they had saved enough to retire on. This is no longer the case' (HKG, 1961).

'[The] pattern of migration deduced from the length of residence table shows that over the last six years the net annual balance of all immigration [that is, both recorded and unrecorded] averaged 60,000' (HKG, 1961).

'The hope of rice for tomorrow' and industrialization

Although the first surge of Chinese immigrants was over in 1950, it turned into a steady inflow at the annual level of 60,000, which was by no means small. At the same time, the fertility level of the immigrants began to rise. It was a young, healthy, and soon to be fertile, immigrant population. It can be seen that fertility did not begin to decline until 1962 (HKG, 1972). It can also be seen that the second upsurge of Chinese immigrants came in 1962, giving the government little breathing space. But before going into these demographic changes, we shall turn to another serious economic and social problem for the immigrants beside housing, namely that of 'the rice bowl'.

Hong Kong was an entrepot before World War II. The People's Republic of China participated in the Korean War almost immediately after its own establishment, and it led to the United States and United Nations embargo on trade with China. Hong Kong's entrepot role died. Industrialization was the only hope. Initially it was entirely an effort of the Chinese immigrants because of their 'hope of rice for tomorrow', but the government was also quick to latch on. Two streams of Chinese immigrants contributed to industrialization, as the government put it: first, the urban Chinese immigrants mainly 'from the North' ('the North' here means mainly Shanghai) who 'brought with them . . . new techniques . . . coupled with a commercial shrewdness and determination superior even to that of the native Cantonese, and . . . new capital seeking employment and security' (HKG, 1957). Second, the rural Chinese immigrants mainly from Guangdong who 'were mostly farmers and crowded in on the town [that is urban Hong Kong and Kowloon], for there alone lay the hope of rice for tomorrow' and 'had to be . . . transformed by some social alchemy from the mentality of the farmer to that of the industrial worker' (HKG, 1957).[3]

'In 1948 there were 1,160 factories and workshops with a total labour force of some 60,000. By mid 1955 there were 2,500 factories and workshops employing 118,000 persons. Another 200,000 persons were employed indirectly or in domestic industries largely in the squatter and resettlement areas. There are now [that is, 1956] 3,319 factories and workshops employing 146,877 persons, and the number of persons directly or indirectly dependent on industry is probably at least 50% of the population' (HKG, 1957). By 1966, the government recognized unmistakably that 'Hong Kong now depends on its industry for its survival, and during the 10 years which have elapsed since 1956, . . . most of our industrialization has taken place . . .' (HKG, 1967).

Different industries – textiles, garment, plastics, toy, electronics[4] – rose and fell throughout the 1950s, 1960s and 1970s. Despite this succession of different industries, the basic character of Hong Kong's industry, that is, labour-intensiveness, remained the same until China began its economic reform in the early 1980s.[5] By then the improving social infrastructure in areas such as education, housing, transport, work remuneration, welfare, lifestyle and leisure had made Hong Kong an increasingly expensive place to hire manpower, resulting in a great differential in labour costs between Hong Kong and mainland China. The 1980s saw a massive removal of Hong Kong factories to Guangdong. By the late 1980s, it was reported that Hong Kong capital employed between 1.5 million to 2.0 million Chinese factory workers in the Pearl River Delta area in Guangdong (KPMG Peat Marwick, 1989). To give a sense of proportion, it should to be remembered that the Hong Kong population was about 6 million at that time; in other words, on the average three to four Hong Kong permanent residents hired one Chinese factory worker. It was clearly more than a relocation of factories; it was in fact an expansion of production capacity into Guangdong. By the early 1990s, manufacturing industries were no longer a major employer of the local labour force. By the mid-1990s, service industries had become the largest employer. The character of the local labour market had changed permanently.[6]

1962: The Second Surge of Chinese Immigrants

'The year (1962) was one in which Hong Kong was subject to almost unprecedented pressure of immigration from China. This flux . . . [was] mostly by clandestine channels . . .' (HKG, 1963). 'For a number of years the problems created by overcrowding in Hong Kong and the impossibility of absorbing immigrants in unlimited numbers had made it necessary to restrict the traditional [free] movement of population between Hong Kong and China. As a result, pressure of entry by clandestine means had for some time posed a serious problem. But in the past this had been largely confined to entry by sea from Macau, and little or no difficulty had been experienced on the land frontier. Toward the end of April however, due . . . to relaxation of control beyond the border [that is, by the Chinese authorities], there was a steady and disturbing increase in attempts, mainly by night, to evade police patrols and penetrate the frontier fence. Over 600 persons were arrested in the border area in the last three days of April and . . . observations showed very large numbers collecting or waiting an opportunity to gain entry. . . . During the early part of May the number arrested mounted alarmingly. . . .

'Much understandable public sympathy for the immigrants was evident in Hong Kong and on several occasions trouble was narrowly avoided in villages near the border to Lo Wu, in order to give them food and clothing, or an opportunity to escape from custody. . . . The influx reached its peak on 23rd May, when 5,620 immigrants were arrested in the frontier area and 5,112 returned to China. On that day it was announced that over 50,000 persons had crossed the border illegally since 1st May. On 26th May the Chinese authorities reinforced normal control measures and the influx ended as suddenly as it had begun. In the course of six weeks over 62,400 persons had been apprehended and returned to China. . . . Although a very large number of immigrants had been intercepted, it was known that many had managed to evade the cordons and check-points and make their way into the urban area. It is now estimated that nearly 60,000 persons succeeded in entering the Colony illegally during May. . . . Although illegal entry across the land frontier had virtually ceased by June, the number of persons entering clandestinely by sea, both direct from China and via Macau, showed substantial increases . . .' (HKG, 1963).

1963–1969: Reduced Immigration from China

The upsurge of Chinese immigrants ended soon. The number of net migrants to Hong Kong returned to normal – with some fluctuation through the 1960s, from a high of 27,800 in 1967 to 1,790 in 1969, with even a negative inflow of 14,400 in 1966. Meanwhile, annual natural population growth declined from 95,500 in 1963 to 63,226 in 1969 (Table 20.1).

Table 20.1 Population growth in Hong Kong, 1962–1969

Year	Population natural growth	Net migrants
1962	91,581	208,500
1963	95,500	20,500
1965	84,574	18,626
1966	77,200	−14,400
1967	74,100	18,300
1968	66,000	27,800
1969	63,226	1,790

Sources: Hong Kong government, 1962, 1963, 1965, 1966, 1967, 1968, 1969.

'The number of persons who succeeded in entering Hong Kong illegally or endeavoured to do so steadily declined during the year.... The main routes of illegal entry into Hong Kong were again Macau by sea and from other small ports nearby on the mainland'. 'The exceptionally large influx of illegal immigrants in 1962 (estimated at 142,000) included many persons with near relatives who had to remain in China. Now that these people have been absorbed as residents of Hong Kong, many are in a position to ask for their wives and children and aged parents to join them . . .' (HKG, 1964). 'Increased vigilance on both sides of the border [between Hong Kong and mainland China] has been a factor in keeping numbers [of illegal immigration] down. There continues to be some illegal traffic to the Colony through Macau' (HKG, 1969).

Social infrastructure building

The second surge of Chinese immigrants from the neighbouring Guangdong province in 1962 was a consequence of the famine after the disastrous Great Leap Forward in China in the late 1950s (Ashton et al., 1984). It might be said that thereafter the 1960s were a peaceful period on the immigration front. But the social problems of the population did not allow any complacency on the part of the government. There were riots in consecutive years, 1966 and 1967. Although the second outbreak was undoubtedly an overspill from the Cultural Revolution, which was raging in the mainland, the first was unmistakably a sign of social discontent among the youth (Commission of Inquiry, 1967). Industrialization on the basis of intensive and cheap labour was not without its social costs, because it was on the one hand dehumanizing and on the other it widened economic and social inequalities until they became glaringly obvious. Mere provision of public rental housing did not solve the problems. Social infrastructure building was brought onto the government's agenda with a change in government attitude and perspective. Before 1956 the government still wondered whether, since 'Hong Kong had granted immediate sanctuary to the refugees . . . was not their rehabilitation and ultimate disposal a matter for some wider organization [alluding to some international agency] rather than a charitable next-door neighbour?' (HKG, 1957). By 1956 the government was asking itself rhetorically: 'By setting itself up as the landlord of some 300,000 refugees, did not Government by that fact alone recognize them as an integral part of the population?' (HKG, 1957). By the mid-1960s, the government was firmly committed 'to [providing] permanently for all the needs of a greatly swollen population – for their employment, shelter, health, education, safety – in short for the creation of a new society' (HKG, 1967). On almost all fronts the government initiated new programmes.

'A seven-year programme of primary school expansion . . . was inaugurated in 1954 with the aim of providing, by 1961, a primary school place for every child of primary school age.... In 1954 the Colony's requirements for technical education were examined, eventually resulting in the building of a new Technical College in Kowloon. In 1954 a tentative experiment in adult education was embarked upon by the government with the organization of classes for factory workers.' 'Concurrently with the expansion of primary education, the supply of trained teachers was increased with the expansion of [teachers' training colleges].... The long-established University of Hong Kong ... [was expanded] ... In October 1963 the new Chinese University was incorporated under ordinance.... The fee income at the two universities covers about 10 per cent of their expenditure, and they therefore rely heavily on government support' (HKG, 1967).

'It was not until 1948 that a Social Welfare sub-section of the Secretariat for Chinese Affairs emerged, and it was not until 1958 that it achieved the status of a department. The swamping of the territory with what the rest of the world termed 'refugees' ... meant that ... the work of the embryonic department was for a long time overmuch associated with the provision of basic relief to the destitute and those made homeless by fire, flood and typhoon.... More recently ... [community development] has shown itself to be of paramount importance to the future of the Colony.... Work in ... [youth welfare] was initially centred on the 8–15 age group, most of whom had little chance of finding school places. But as the problems changed, the approach widened and energy is now [that is, in 1956 during which there was a riot] being turned to the highly constructive work of creating outlets for the more inflammable 14–21 year age group, which finds little of positive consequence to provide diversion from the tedium engendered by a crowded environment' (HKG, 1967). 'In 1957, the first long-term outline plans for the development of medical and health services were prepared.... When the results of the 1961 census showed a need for further re-thinking ... a wholly revised development plan for 1963–72 was proposed ... The main aims of the 1963–72 plan were to institute a network of clinics supported by specialist and large hospital facilities, increase the number of hospital beds per thousand of the population from 2.91 in 1963 to 4.25 in 1972, and maintain and increase the preventive coverage and defence' (HKG, 1967). 'There obviously remain daunting burdens to be shouldered, the achievements of the past [that is, 1956–1966] must however give some feeling of confidence in searching the horizon.... Not even the sharpest critic will deny that progress has been made in almost every field and substantial progress in some, that if life is still hard and difficult for many it is also, for many, less grinding and more tinged with hope than it was' (HKG, 1967).

1970–1979: Inflow of Vietnamese Refugees, and Fertility Decline

'The growth in the past few years has not been as fast as in the early 1960s and this can be attributed chiefly to the continued decline in the number of births since 1962. The crude birth rate dropped from 35 births per 1,000 of the population in 1962 to 19.7 in 1971. . . . Approximately 55 per cent of the urban population is now of Hong Kong birth.' 'Applications for entry by close relatives of residents and by former residents were dealt with liberally. . . . Hong Kong continued to attract illegal immigrants, mainly from China, Macau, and the countries of South-East Asia. . . . There was a large increase in illegal entry from Macau and China and a noticeable rise in the number of travellers from Vietnam and the Khmer Republic who entered Hong Kong ostensibly in transit but remained illegally' (HKG, 1972).

'Hong Kong's rate of natural increase in population has dropped significantly in the past few years. Natural population increase was 14.7 per thousand in 1973, as against 27.5 in 1963. Births dropped from . . . 33.5 per thousand . . . 10 years ago to . . . 19.8 per thousand . . . in 1973.' 'During the year 21,738 cases of illegal entry were recorded . . . These figures do not tell the full story, as many illegal immigrants are not detected until long after their arrival, by which time it is often too late to attempt repatriation' (HKG, 1974).

'The average annual rate of increase over the 10-year period [that is, 1966 to 1975] was two per cent. The rate year by year fluctuated owing to changes in migration flow. But the rate of natural increase dropped steadily over the period from 23.1 to 13.3 per thousand. This was the result of a decline in the birth rate from 28.1 per thousand in 1965 to 18.3 per thousand in 1975, with the death rate remaining stable at about five per thousand . . .

'The population of Hong Kong is still a very young one – more than 43 per cent of the population in 1975 was below the age of 20. But the median age of the population – which 10 years ago was 20.8 – is now 23.3 years. The proportions between the different sections of the population have also changed considerably. In 1965, 40.8 per cent of the population was under 15; now it is 31.4 per cent. The relative figure for those aged 65 and over has risen from 3.6 per cent to 5.5 per cent. This indicates that there is a greater potentially productive population (aged 15–64) available to support the infants, those who are being educated, or those who have retired. The dependency ratio – the ratio of the young and the retired to all those in the 15–64 age group – dropped from 800 per thousand in 1965 to 584 per thousand in 1975.

'The low proportion of the population in the under 15 age group is the result of a decline in the birth rate – which is low even compared with some developed countries. . . . This decline in the birth rate is partly the result of women having fewer children, as well as a decrease within the prime childbearing age groups in both the number of women and the proportion of currently married women. In recent years, later marriages have also contributed, while improvements in education and job opportunities for women have almost certainly played their part. . . .

'There was a general decline in mortality after 1951. The death rate dropped to the level of about five per thousand in 1964; since then it has remained much the same. The average life-span of both males and females has increased by 7 per cent over the past 15 years, but male and female expectations of life at birth are still very different. Females born in 1975 should live, on average, 7.56 years longer than males; their expectation of life at birth was 75.5 years and 67.94 years respectively' (HKG, 1976).

'In 1975 some 11,588 of the people registering for identity cards claimed to have entered Hong Kong illegally at some time in the past. During the same year 2,443 illegal immigrants were removed from Hong Kong, including 1,133 who were returned to China under the arrangements introduced in November 1974. . . . In May 1975, Hong Kong gave temporary asylum to 3,900 Vietnamese refugees who had escaped from the fighting in and around Saigon on a small ocean-going vessel . . .' (HKG, 1976).

'During 1976 the government completed its first overall review of its policies which related to the growth of Hong Kong's population. As a result, the government announced in November that existing controls over legal and illegal immigration are to be maintained. Efforts will also be continued by the Medical and Health Department and the Family Planning Association to provide yet more comprehensive family planning services. Although Hong Kong's birth rate fell . . . in the past 10 years, it is calculated that the number of women in the fertile age group between 20 and 35 will grow from 478,400 in 1976 to 738,700 by 1986' (HKG, 1977).

'The aftermath of the political changes that took place in Indo-China in 1975 continued to affect Hong Kong. . . . [More refugees arrived.] . . . Apart from the refugees, more than 5,200 illegal immigrants from Vietnam were permitted to stay in Hong Kong after they had surfaced by registering for identity cards. This led to a sharp increase in applications for the entry of dependants from Vietnam. . . . Immigration from China has declined . . . to 27,599 in 1976 . . . The figures include illegal immigrants whose numbers are

Table 20.2 Population changes in Hong Kong in the 1970s

Year	Population natural growth	Birth rate (per 1,000)	Number of illegal migrants
1970	59,136	20.0	—
1971	59,415	19.7	—
1972	58,212	19.7	17,271
1973	14.7 per 1,000	19.8	21,738
1974	14.5 per 1,000	19.7	22,928
1975	13.3 per 1,000	18.3	—
1976	12.6 per 1,000	17.7	—
1977	55,000	—	—
1979	11.7 per 1,000	16.9	—

Sources: Hong Kong Government, 1971, 1972, 1973, 1974, 1975, 1976, 1977, 1978, 1979, 1980.

impossible to estimate exactly.... More than 200,000 immigrants from China and Vietnam have been resettled [in Hong Kong] since 1971 ... (HKG, 1977). 'In the first half of the 10-year period, the decline in the birth rate was caused by there being fewer women in the prime child-bearing ages of 25–34, and by women generally having fewer children. In the second half, the decrease was mainly the result of fewer births. In recent years, later marriages also have contributed to this trend, along with improvements in education and job opportunities ...' (HKG, 1980). 'The events of 1979 which had the most dramatic impact on immigration control were the big increases in the numbers of Vietnamese refugees arriving in Hong Kong and the high level of immigration (both legal and illegal) from China' (HKG, 1980).

Fertility decline and family planning

Although much noise was made by the government and the media about the influx of Vietnamese refugees, they did not make a permanent mark on the Hong Kong population as most of them emigrated elsewhere or were repatriated back to Vietnam over the next three decades. Throughout the 1970s legal and illegal immigration from the mainland continued on a more modest scale, but industrialization and social infrastructure building managed to buffer its impact and the Chinese immigrants were absorbed into the local community. The immigration front continued to be relatively quiet during the 1970s until 1980, when the third upsurge in Chinese immigration took place. What was far more important, and escaped public notice, was that by the 1970s one of most significant demographic transitions, namely, the fertility decline, had already occurred quietly. It emerged in around 1962 and by 1971 it was certainly permanent.

The abject poverty of the early 1950s, together with the high fertility of the first influx of Chinese immigrants who were mostly young and fertile, prompted the setting up of the Family Planning Association of Hong Kong (FPAHK) by a small group of wealthy, educated and socially conscious volunteers in 1950. The FPAHK offered contraceptive education, publicity and clinical services to all women who needed them. Its service to the Chinese immigrants can hardly be ignored since the Hong Kong Government did not offer any contraceptive services until the early 1970s.[7] The current contraceptive practice rate for married women aged 14–55 rose to 45 per cent by 1967 and has risen to higher levels ever since. The total fertility rate has been decreasing – initially rapidly and later more gradually – reaching 2.5 in 1976, falling further to 1.5 in 1985, and stabilizing at the low level of 1.2 in 1993. In many ways, the early success of family planning in Hong Kong underpinned many of the more visible successes in later years because it made wealth accumulation at a fast rate and social infrastructure building at a more leisurely pace possible. It is most notable and perhaps also most exceptional that the family-planning movement was a non-governmental initiative and that contraception was and still is accepted by the women or couples concerned without any coercion.

Various explanations have been offered for the decline of fertility in Hong Kong, and all are plausible. Macrosociological ideas of industrialization, urbanization and modernization were called upon in the explanations, and so were microsocial ideas of changing rationalities, values and attitudes towards child-bearing and contraception.[8] Buried under these explanations is one fundamental demographic fact, namely, the bulk of women in reproductive age changed from immigrants to local-borns in the early 1970s.[9] In other words, the burden of fertility shifted to a new generation of young women who were raised in a social and economic environment different from that of their older immigrant counterparts; they benefited from widening opportunities for education and employment as Hong Kong improved.[10] The fertility decline is nevertheless confirmed for the future. It synchronizes with other concomitant socio-demographic trends, namely, the high rate of female labour force participation, the delayed timing of first marriage for both sexes, and the high prevalence of contraceptive use for marital sex. All these trends are indicative of an economic, social and cultural configuration – a structural one – which sets a limit to fertility.[11] In this sense, Hong Kong has already converged with the model of a modern society as exemplified by many developed countries.

1980: The Third Surge of Chinese Immigrants

'The life expectancy for those born in 1980 is 70.1 years for males and 76.8 pears for females. . . . About 57 per cent of the population was born in Hong Kong.' 'Reflecting the baby boom of the 1950s and early 1960s, the number of women in the fertile age group of 25 to 34 years will increase substantially from 387,800 in 1980 to 538,400 by 1990' (HKG, 1981). 'In 1980 the problems created by the massive inflow of illegal immigrants from China overshadowed all other immigration issues and showed no sign of easing. Therefore, . . . the government made a major change in its policy. From October 23, no illegal immigrant from China was to be allowed to remain in Hong Kong. Those already here were given three days in which to register for an identity card and new legislation was enacted to deter others from coming. . . . Prior to this, the last change in the policy had been in 1974, when the practice of allowing all immigrants from China to remain was ended; from then on, those arrested on arrival were repatriated. However, all others who evaded capture and subsequently 'reached base', that is, gained a home with relatives or otherwise found proper accommodation, were permitted to stay. . . . In the following three years, the implementation of this dual policy brought no major difficulties – about 6,000 illegal immigrants from China were reaching base every year and were being absorbed without strain. But 1978 saw a change. With the new, more liberal environment in China, the greater freedom of movement there and the increase in contacts with the rest of the world, those living in the communes were not only able to see the attractions of Hong Kong more clearly but also found it easier to reach the border. The rise in numbers entering the territory was dramatic: [In 1977, 1,800 were arrested on arrival and repatriated and 6,600 evaded capture and remained (estimated); in 1978, 8,200 and 28,100 respectively; in 1979, 89,900 and 107,700 respectively; in 1980 (January to October), 80,500 and 69,500 respectively.] . . . The situation was aggravated by the concurrent steady flow of some 55,000 legal arrivals a year from China, most of whom had been given permission by the Chinese authorities either to visit Hong Kong or to pass through in transit to other countries, but who remained in the territory permanently.' 'Regulations [were introduced] in September making it compulsory to carry identity cards in the New Territories . . . [and were] later extended to the whole territory . . . in late October . . .' (HKG, 1981).

Sealing Hong Kong off into a separate Chinese population

A fixed quota of immigration from mainland China was negotiated between the Chinese government and the Hong Kong government subsequently.[12] Mainly those in mainland China who were connected to Hong Kong by marriage or by family were admitted. The annual quota was below 30,000, that is, an annual increase of less than 5 per cent of the base population. It was a leisurely paced economic, social and cultural absorption of new immigrants in the local community – a situation very different from that of the 1950s and 1960s when the local community was overwhelmed by new immigrants. A distinctive economic, social and cultural life and tradition had been developing. In this sense, Hong Kong had become a separate Chinese population. It was not envisaged that the return of sovereignty in 1997 would affect this separation from mainland China to an appreciable extent in the foreseeable future. It seemed safe, at least for the meantime, to conclude that the chapter of immigration from mainland China was closed.

Historically, ethnic non-Chinese were never a significant proportion of the Hong Kong population. The very great majority of them were migrant workers – whether colonial officers and traders in the old days or professionals in more recent times – and their dependants. Very few of them could be considered as immigrants in the sense that they wished to settle down permanently in the country. It is safe to say that Hong Kong has always been a Chinese population historically, even though it was a British colony for about a century. At the present, the largest non-Chinese ethnic group is the Filipinos.[13] There were 157,026 Filipino domestic helpers in Hong Kong in 1996 (HKG, 1996). Hong Kong laws prohibited them from becoming permanent residents.

The ethnic 'Chineseness' of the Hong Kong population is beyond doubt although a very substantial proportion of the ethnic Chinese permanent residents hold non-Chinese passports, which is mainly a legacy of the

British rule, including both the status of British National (Overseas) of all locally born ethnic Chinese and the status of British Citizen granted to some 40,000 families as a contingency measure to retain the professionals and key role players in Hong Kong after the political disturbances in Beijing in 1989. There was, consequently, an impressive emigration, a mini-diaspora, during the political uncertainty in expectation of the change of sovereignty.

1981–1988: Maturing Demographically

1981.   'One fact that the 1981 Census clearly establishes is that in age structure Hong Kong is not getting any younger. There has been a continuous decline in the proportion of young people aged 0–14, while old people aged 65 years and over have slowly risen since the 1971 Census from 4.5 per cent of the population to 6.6 per cent. The proportion of the working age population, from 15 to 64, has increased from 59.7 per cent to 68.6 per cent over the past 10 years' (HKG, 1982).

'As a result of the abolition of the "reached-base" policy – which allowed illegal immigrants from China who reached the urban areas to remain – in October 1980, coupled with a new regulation which makes it compulsory for everyone over 15 years of age to carry a legal form of identity, there was a substantial reduction in the level of illegal immigration during 1981' (HKG, 1982).

During the 1980s, the average annual growth rate declined continuously from 1.6 per cent in 1981, down to 1.37 per cent in 1988. Meanwhile, the rate of natural increase dropped from 12 per thousand to around 8 per thousand in the late 1980s. This was the result of the birth rate declining from 17 per thousand in 1982 to about 13 per thousand, and the death rate remaining stable at about five per thousand. The inflow of legal immigrants from the mainland settling in Hong Kong remained steady at about 27,000 over the 1980s. Although illegal immigration still took place, it was on a much reduced scale.

Mortality, population ageing, household resources and social mobility

In comparison with other demographic aspects, mortality has been most uneventful. The Hong Kong population has not suffered any serious epidemic or famine since World War II.[14] Infectious diseases are not the major causes of death. Life expectancy has improved slowly over the last few decades to a level comparable with that of developed countries. Also, as in developed countries, neoplasm and heart and circulatory system diseases are increasingly dominant as causes of death, indicating that rising affluence has begun to take its toll.[15]

More significant is the gradual ageing of the population, which is clearly indicated in the moving population pyramid across the series of censuses.

Table 20.3 Major population indicators for the 1980s

Years	Annual growth rate (%)	Natural growth rate per 1,000	Migrants from mainland
1981–2	1.6	12	—
1982–3	1.4	11	27,000
1983–4	—	10	27,700
1984–5	—	9	27,300
1985–6	—	8	27,100
1986–7	1.27	—	27,300
1987–8	1.37	—	28,000

Sources: Hong Kong Government, 1982, 1983, 1984, 1985, 1986, 1987, 1988, 1989.

Hong Kong is now one of the three most aged Asian regions, the other two being Japan and Singapore (Phillips and Bartlett, 1995). There are of course some worries about the long-term economic and social consequences of population ageing (Chow, 1993; MacPherson, 1993; Kwan and Davis, 1994; Ngan and Wong, 1994; Bartlett and Phillips, 1995). However the likelihood and the extent of such consequences need to be assessed at the family or household level since Hong Kong society is largely organized with individual families as building blocks. Hong Kong society, that is, is mainly a collectivity of families. Also, since the great majority of households are families, the household and the family are roughly exchangeable concepts for our analytic purpose.[16] The key concept we use for our analysis is household resource. What is ascertainable from census data is that the average household and hence the family has been becoming more resourceful over the last four decades. As regards human capital, in 1991 the proportion of households having at least one member working stood at a healthy level of 32.5 per cent and that of households having at least one member educated to Form Five or above had increased to 32.7 per cent (Lui, 1996). Nine-year compulsory education from 1980 and a recent expansion of tertiary education mean that almost all adults will be at least literate and about one-sixth of them will be university-trained in the future. It implies that most families in the long run will rely on white-collar breadwinners if the local economy continues to move towards the service industries. There are no strict retirement regulations in Hong Kong and white-collar workers can work into their old age as long as their health permits and as long as the local labour market does not slacken off. The impact of population ageing will perhaps be less severe than is usually feared.

Household resource is closely related to another societal phenomenon, namely social mobility. Social mobility, in the sense of inter-generational change in occupation, is impressively in evidence.[17] For example, less than one-quarter of the fathers of the incoming batch of students at one of the local universities in 1988 were professional, administrative and managerial workers, while these students were expected to belong to this occupational category in the future (Hong Kong Polytechnic, 1989). As mentioned earlier, the local labour market has changed as factories are moved to mainland China. Factory jobs have vanished. At the same time, the removal of factories is in fact an economic expansion, insofar as supervisory and managerial jobs, and jobs in the supporting tertiary industries have been being created. These new jobs require more years of education and training. At the same time, improving educational opportunities produces the necessary labour power from the younger generation. It results in upward mobility for a substantial number of local families. From a sociological perspective, the social stratification into housing classes mentioned above, and the social mobility in the form of inter-generational occupational mobility, pose an enthralling sociological aporia.

Nuptiality and sexuality

Beginning in the 1970s, the long-term social and cultural consequences of economic structural and social infrastructural changes began to emerge. They were particularly striking in nuptiality and sexuality.[18] Although the proportion of women never married at the end of their reproductive years (age 44) has not increased significantly, the median age at first marriage for Hong Kong women has been increasing since the 1970s[19]. At the same time, the divorce rate is increasing. Marriage as a social institution seems to continue to be valued and yet marital life is shorter and is more susceptible to disruption for the younger cohorts of women than for their older counterparts.[20] As every marriage involves a man and a woman, what is observed in women holds generally for men. In sex, however, men and women differ. While unmarried men's involvement in premarital sex is traditionally tolerated by parents, unmarried women's involvement was seriously frowned upon until recently. The traditional 'arithmetic' is that premarital sex means premarital loss of virginity which is a serious matter for an unmarried woman (but not for an unmarried man) and the man involved is the 'culprit' who 'steals' the woman's virginity. Marriage with the 'culprit' was often the solution sought by the 'victim' or her parents to premarital pregnancy. Had this traditional solution continued, the rising age at first marriage would mean the decline of premarital sex. But premarital sex is known to occur fairly often. Empirical data are scarce, but at least one study in 1986 found that slightly less than one-fifth of sampled women and slightly less than one quarter of sampled men aged 18–27 reported their involvement in premarital sex.[21] The co-occurrence of delaying first marriage and a rising incidence of premarital sex is made possible by the acceptance of induced abortion as a more preferred solution than marriage to premarital pregnancy.[22] Tradition has changed, doubtless.[23]

The rising divorce rate may be another consequence of the changing tradition. The use of commercial sex by men, both married and unmarried, is traditionally tolerated, while its use by women is not. Husbands' use of commercial sex is traditionally tolerated by their spouses. In a similar vein, husbands' involvement in non-commercial extramarital sex is traditionally more tolerated by their wives than vice-versa. This traditional value is clearly in conflict with the idea of sexual equality which is part of the idea of modernity.[24] Naturally, acceptance of the idea of sexual equality is higher among women than among men.[25] The ideational conflict can become manifest in two argumentative forms. Either the wife concerned pursues sexual equality by demanding that the husband concerned refrain from promiscuity, which is refused. Or the wife concerned pursues sexual equality by becoming equally promiscuous herself, which is not tolerated by the husband concerned. In either case, divorce may ensue (Luis, 1997). The nuptiality change and the concomitant and closely related sexuality change indicate a more fundamental social and cultural change, namely, that the Hong Kong population is reproducing and structuring itself into a separate culture, one that is separately identifiable from the mainland culture. This change, since it is a summative phenomenon arising out of the behaviour of individuals, is a consequence of the natural renewal of age cohorts as social and cultural actors, with the older cohorts phasing out and the younger cohorts phasing in. More importantly, this change is self-generative within the local population, and, unless there is another very massive immigration from outside the local population, will change in the future according to its own cultural dynamics. In this respect, Hong Kong has successfully formed and will maintain its own cultural autonomy.[26]

1989–1996: A Wave of Emigration

'The subject of emigration began to attract media and public attention in late 1987 when there were signs that more people were leaving Hong Kong. . . . [It was estimated] that 30,000 persons emigrated in 1987 and 45,800 in 1988 compared to a historical average in the early 1980s of around 20,000. The forecast for 1989 was 42,000. The reason for the increase was a combination of factors; on the one hand some people were nervous about Hong Kong's future after 1997 under Chinese sovereignty, while at the same time there were more immigration opportunities available in the more popular destination countries. . . . Research also revealed that, while a disproportionate number of emigrants were well educated and professionally skilled, the total number of such people in the community was continuing to rise' (HKG, 1990). The trend was maintained over the period. The estimate of the number of persons emigrating was around 60,000 for the years 1990–94, and then dropped to 43,100 in 1995. On the other hand, the number of new arrivals from the mainland increased steadily from 28,000 to 45,986 for the same time period (HKG, 1991, 1992, 1993, 1994, 1995, 1996). An

Table 20.4   Indicators of migration in Hong Kong, 1989–95

Year	Number of emigrants	Number of immigrants from mainland			
		Total	Wives of local residents	Children of local residents	Husbands of local residents
1989	42,000	27,300	9,565	13,216	937
1990	62,000	28,000	10,302	13,259	1,042
1991	60,000	26,800	10,113	12,513	1,020
1992	60,000	28,400	11,128	12,457	1,082
1993	54,000	32,900	—	—	—
1994	62,000	38,200	—	—	—
1995	43,100	45,986	18,274	23,033	1,572

Sources: Hong Kong Government, 1990, 1991, 1992, 1993, 1994, 1995, 1996.

overwhelming majority of those newcomers were wives and children of local residents (table 20.4).

In 1995, '[there were] 38,200 more births than deaths and a net inflow of 120,600 people. . . . The rate of natural increase dropped . . . to six per 1000.' 'It is estimated about . . . 43,100 [people emigrated] in 1995. . . . Of those . . . about 15,700 were managers, administrators, professionals and associated professionals' (HKG, 1996). 'The one-way permit quota agreed upon with the Chinese Government for legal immigration from China was increased from 105 to 150 a day from July 1, 1995.

Emigration and family strategy

Through the 1960s, the 1970s and the 1980s, there was always some outflow of Hong Kong residents to other countries (exact figures are not known), but the only significant wave of emigration from Hong Kong after the 1950s was during the uncertain period when the British and Chinese governments negotiated the future of the island. The turmoil in Beijing in the summer of 1989 saw the climax of emigration in intent if not in action. When the political climate stabilized in the mid-1990s, return migration began. There is no accurate assessment of net emigration, and in fact such assessment is not possible given the often less than complete severance of emigrants from Hong Kong. Clearly some kinds of family strategies are employed to maintain close social and economic connections with, if not a physical presence in, Hong Kong. Many emigrants split their families into two 'parts' with one in Hong Kong and the other in their adopted countries, travel frequently between Hong Kong and their adopted countries, or continue to make their living in Hong Kong. In any case, unless an emigrant declares that he or she is giving up Hong Kong permanent resident status, by law he or she is entitled to land and work in Hong Kong indefinitely. The social implications are nevertheless clear: the Hong Kong population as a social network has been increasingly internationalized.

Conclusion: Future Prospects

One thing is certain: it is becoming increasingly difficult to apply the conventional demographic and sociological concept of a population as a human collectivity who live, work, play, reproduce and die at a specific geographical location to the Hong Kong population.[27] The international-ization of the Hong Kong population is one of the factors, but there are some more important ones. The removal of factories to mainland China implies that an unknown but definitely substantial proportion of the labour force work there or commute frequently between their homes in Hong Kong and their workplaces in mainland China. Some of them establish a second household there. In this sense they are migrant workers. This trend is expected to grow since further economic integration of Hong Kong into China is expected.

The neighbouring special economic zones of Shenzhen, Zhuhai and Shekou are eager to connect themselves more closely with Hong Kong by attempts to build roads and bridges to Hong Kong, to open their border checkpoints twenty four hours a day, or even to encourage Hong Kong residents to establish homes there and commute to their workplace in Hong Kong daily.[28] The cost of accommodation and living is substantially lower there. Even without the encouragement of the local governments, because of their proximity these zones and the neighbouring towns outside them have already become places of recreational and leisure activities for Hong Kong people.[29] The differential in cost of living has already prompted some retirees, elderly people or people of little means to move their residence to mainland China voluntarily and visit their home in Hong Kong only occa-sionally or when there is a need.[30] Hong Kong continues to provide them with medical and social benefits, which are their entitlements as Hong Kong permanent residents. The government is revising its social welfare policy to suit this trend.[31] There is no population policy in Hong Kong, and we do not expect one even in the future. But other polices are clearly in need of rethinking. The tax base is shifting unfavourably as more and more eco-nomic activities move outside Hong Kong, but the improving social infrastructure is constantly demanding heavier and heavier financial com-mitment from the government coffers. Since permanent residence confers entitlements and entitlements are costly to the government, the obligations of permanent residents need to be defined. It will lead to a redefinition of permanent residence, which will in time generalize to a fundamental demo-graphic question: what then is a population?

Acknowledgment

Mr Ho Wing-chung, a postgraduate student of mine, has helped in compiling the references and notes.

Notes

1. The description given here is a non-Marxist one. For radical Marxist ones, see, for example, Lai (1995). For a conventional sociological conception of the formation of classes in South Korea, Taiwan, Hong Kong and Singapore, see, for example, Koo and Hsaio (1995).
2. Some local researchers take a different view on whether the social infrastructure has really been improving. They argue that, despite economic growth, about one third of the Hong Kong population still live in absolute poverty, with millions of others suffering a rapidly increasing relative poverty. See, for example, Macpherson (1992).
3. We focus on the connection between immigration and industrial change. For an examination of the causes and implications of rapid industrial growth in Hong Kong from the early 1950s to the late 1980s, see Henderson (1989).
4. The electronics industry in Hong Kong was supported by foreign investment, notably from the United States. This has prompted some Marxist analyses, for example, Djao (1978). Even then, the great majority of factories at that time and in subsequent decades are believe to have been owned by Hong Kong Chinese.
5. Obviously an alternative description can be made in terms of factors of production: see, for example, Schiffer (1991).
6. Ng (1997) describes the situation as follows: 'Since the mid-1980s, largely as a result of the opening up of the People's Republic of China, Hong Kong has experienced massive economic restructuring, involving massive relocation of manufacturing plants, downsizing, and outright closure. The garments and electronics factories have been hardest hit, industries whose workforces are composed mainly of middle-aged females working in semiskilled jobs. . . . Several ways of dealing with the . . . [resulting] hardship are . . . : some stayed out in the dwindling economic plants and some found new service sector jobs; . . . some coped by relying on their family of origin to absorb the hardship, while others withdrew to become full-time houseworkers; some tried to start their own business.
7. There was a debate between Wat and Hodge (1972) and Freedman (1973) on whether the family-planning programme was the main of fertility decline in Hong Kong. We tend to agree with Freedman who argued that, although fertility decline was bound to happen sooner or later in a rapidly developing society and economy such as Hong Kong, an organized programme spreading birth-control information and devices would affect the timing of the decline by serving as an exogenous variable.
8. For a macrosociological explanation, see, for example, Leete (1994). For a microsociological explanation, see, for example, Family Planning Association of Hong Kong (1984, 1989).
9. By-censuses showed that the proportion of women aged 20–34 who had been born in Hong Kong rose from 30.2% in 1966 to 61.4% in 1976. We use an age

range narrower than the usual convention (that is, ages 15–44) because very few births were to women outside this shorter range.

10. The concomitant changes in the roles of women are discussed by several researchers, for example, Kao (1987) and Ma (1991).

11. Shek (1996) observes in his sample of parents that the value of children to Hong Kong Chinese parents has changed. Although the respondents generally agreed that children contribute to parents' own personal growth, spousal relations, family happiness and family wholeness, a majority also perceived that parenthood was associated with increased burdens and personal sacrifice. These parents did not seem to emphasize strongly the traditional Chinese values of security and posterity. Fathers tended to perceive parenthood as more positive and less burdensome than did mothers.

12. Legal provision and policy implementation were discussed by some researchers, for example, Lui (1983) and Vagg (1993).

13. Many educated married women are released from their households to work outside the home because of these migrant workers. They are now part of the regular scene in middle- to upper-income households. See Ozeki (1995).

14. Goldman (1990) suggests that mortality schedules in Taiwan, Hong Kong, Singapore and Korea have been characterized by excessively high death rates of older adult males. He goes on to mention that this excessive mortality has progressively diminished over the years and that by the late 1970s death rates show only slight deviations from the western model of life tables. He suspects that it was a result of exposure to high levels of tuberculosis in the past.

15. Rising affluence is partially reflected in individual consumption across the decades. Individual consumption covers goods and services which are actually consumed by households (e.g. education services), irrespective of whether the ultimate bearers of the costs are household themselves (private tutors), private non-profit bodies or the government. Individual consumption per capita rose continuously from 1991 to 1996: HK$1,831 in 1961, HK$2,686 in 1966, HK$4,463 in 1971, HK$8,437 in 1976, HK$21,013 in 1981, HK$36,216 in 1986, HK$73,540 in 1991, and HK$122,007 in 1996, at current market prices. In spite of inflation, there were still substantial net improvements.

16. For example, in the 1986 and 1996 by-censuses, nearly four-fifths of the population fell into one of the four categories of households, i.e. one unextended nuclear family, one vertically extended nuclear family, one horizontally extended family, and two or more nuclear families. We are not concerned with the 'nature' of the Hong Kong family. Interested readers are referred to Lau's (1981) idea of utilitarian familism. In the 1970s there were numerous publications about changes in kinship, family networks, family structures, family roles and family norms – for example Hong (1972, 1973), Mitchell (1973), Podmore and Chaney (1974), Wong (1975, 1977), Salaff (1976a) and Rosen (1978). The research topic went out of favour in the late 1980s.

17. Post (1996) concludes on the basis of 1981 and 1991 census data that family background and gender have played smaller roles in determining which children go to secondary school and there was no concomitant increase in social selection at the post-secondary level. The earlier Post (1994), based on 1976, 1981 and 1986 census data, arrives at similar conclusions. The even earlier Pong and Post (1991) concludes differently as regards social selection for higher educational opportunity, but their other findings agree with Post (1994, 1996).

The very early Podmore and Chaney (1973) suggested that Hong Kong young adults accepted parental authority in matters relating to occupational choice and attempted to explain it in terms of geographical limitations, the local political features, and the importance of family in finding jobs. They proved to be mistaken in the emphasis of their study.

18. I have chosen not to adopt the conventional sociological focus on social and cultural change in the family because it will deviate from social demography too far. For a conventional sociological view of the Hong Kong based on the interplay of Chinese traditionalism and western modernism, see, for example, Chan and Lee (1995). A non-demographic description of the renewal of the Chinese family can be made in terms of filial piety and family cohesion, see Cheung et al. (1994). In a different context of discussion, Wong (1995) suggests that family firms were prevalent among privately owned Chinese enterprises in Hong Kong.

19. Salaff (1976b) analysed the reasons for delaying marriage among a sample of young women in the early 1970s. Luis and Chan (1993) analysed the determinants of the transition to marriage among young Hong Kong women in the mid-1980s.

20. An early work on attitudes towards marriage and the family in the early 1970s is Podmore and Chaney (1972). They argued that Hong Kong society would converge to the norm of industrial societies despite the underlying cultural differences.

21. The Family Planning Association of Hong Kong (no date), p. 127.

22. Service statistics of the Family Planning Association on induced abortions in unmarried women indicate this. See the Association's annual reports.

23. In a study in the early 1970s, Raschke (1976) ascertained that Hong Kong Chinese college students were less sexually permissive and less sexually active than their United States midwest counterparts. This contrasts sharply with findings from studies in the early 1990s.

24. The opening up of mainland China has increased the opportunities for Hong Kong men to be involved in extramarital sex there. For Hong Kong men's own narratives about their extramarital activities in mainland China, see Young and Kwan (1995). Action against this by the women affected and by feminists is increasing. See, for example, Kwok (1995).

25. When respondents in a local study in 1986 were asked whether men and women were equal in sexual rights, that is, whether a woman is free to engage in any sexual relationship which is permitted to a man, 67.5% of the women sampled agreed, whereas only 53.7% of the men did. See the Family Planning Association of Hong Kong (no date), p. 112.

26. This broad claim of ours should perhaps be justified with evidence on more cultural aspects than nuptiality and sexuality. There has always been a school of researchers, especially anthropologists, who emphasize cultural continuities from mainland China to Hong Kong among immigrants and their children, see, for example, Hsieh (1985). Our claim in fact does not contradict theirs, and the difference is that we look forward while they look backward.

27. Roland Pressat defines a population as '[a] group of individuals coexisting at a given moment and defined according to various criteria. The term population usually denotes all the inhabitants of a specific area (state, province, city.

etc.) . . . In an even more restricted sense the term is used to refer to any group under study . . . where entry into and exit from the population can be seen as determining its size and structure, in the same way that birth, death and migration affect the population at large.' See Wilson (1985), p. 176.

28. For examples, see the *Hong Kong Economic Journal* (a local daily) of 3 March 1997, *Ming Pao* (a local daily) of 24 April 1997, the *Hong Kong Standard* (a local daily) of 3 June 1997, and Ma (1996).

29. For example, *Apple Daily*, the most popular newspaper, has a regular column surveying and introducing brothels in Shenzhen and neighbouring towns across the border, as a 'guide' for brothel visitors. Brothel visiting is undoubtedly a recreational and leisure activity. This column is simply meeting a community need.

30. For example, a local non-governmental organization called Helping Hands is building an infirmary in Guangdong for the Hong Kong elderly.

31. For example, the government is considering a change in policy to allow eligible Hong Kong elderly to receive Comprehensive Social Security Assistance (CCSA) even if they reside in Guangdong permanently.

References

Ashton, Basil, Hill, Kenneth, Piazza, Alan and Zeitz, Robin (1984), Famine in China, 1958–61, *Population and Development Review*, 10 (4), 613–45.

Bartlett, Helen P. and Phillips, David R. (1995), Ageing trends – Hong Kong, *Journal of Cross-Cultural Gerontology*, 10 (3), 257–65.

Chan, Hoi-man and Lee, Rance P. L. (1995), Hong Kong families: at the crossroads of modernism and traditionalism, *Journal of Comparative Family Studies*, 26 (1), 83–99.

Cheung, Chau-kiu, Lee, Jik-joen and Chan, Cheung-ming (1994), Explicating filial piety in relation to family cohesion, *Journal of Social Behavior and Personality*, 9 (3), 565–80.

Chow, Nelson (1993), The changing responsibilities of the state and family toward elders in Hong Kong, *Journal of Aging and Social Policy*, 5 (1–2), 111–26.

Commision of Inquiry (1967), Kowloon disturbances 1966, *Report of Commission of Inquiry*, Hong Kong: Government Press.

Djao, A. Wei (1978), Dependent development and social control: labour-intensive industrialization in Hong Kong, *Social Praxis*, 5 (3–4), 275–93.

Freedman, Ronald (1973), A comment on 'social and economic factors in Hong Kong's Fertility Decline' by Sui-Ying Wat and R. W. Hodge, *Population Studies*, 27 (3), 589–95.

Family Planning Association of Hong Kong, The (1984), *Report on the Survey of Family Planning Knowledge, Attitude and Practice in Hong Kong 1982*. Hong Kong: The Family Planning Association of Hong Kong.

Family Planning Association of Hong Kong, The (1989), *Report on the Survey of Family Planning Knowledge, Attitude and Practice in Hong Kong 1987*. Hong Kong: The Family-Planning Association of Hong Kong.

Family Planning Association of Hong Kong, The (no date): *Working Report on Adolescent Sexuality Study 1986*. Hong Kong: The Family Planning Association of Hong Kong.

Goldman, Noreen (1980), Far Eastern patterns of mortality, *Population Studies*, 34 (1), 5–19.

Henderson, Jeffrey (1989), Labour and state policy in the technological development of the Hong Kong electronics industry, *Labour and Society*, 14, supplement, 103–26.

Hong, Lawrence K. (1972), A comparative analysis of extended kin visitations, cohabitations, and anomia in rural and urban Hong Kong, *Sociology and Social Research*, 57 (1), 43–54.

Hong, Lawrence K. (1973), A profile analysis of the Chinese family in an urban industrialized setting, *International Journal of Sociology of the Family*, 3 (1), 1–9.

Hong Kong Government (1959), *Hong Kong Annual Report 1958*. Hong Kong: Government Press.

Hong Kong Government (1960), *Hong Kong Annual Report 1959*. Hong Kong: Government Press.

Hong Kong Government (1961), *Hong Kong Report for the Year 1960*. Hong Kong: Government Press.

Hong Kong Government (1962), *Hong Kong Report for the Year 1961*. Hong Kong: Government Press.

Hong Kong Government (1963), *Hong Kong 1962*. Hong Kong: Government Press.

Hong Kong Government (1964), *Hong Kong 1963*. Hong Kong: Government Press.

Hong Kong Government (1965), *Hong Kong 1964*. Hong Kong: Government Press.

Hong Kong Government (1966), *Hong Kong 1965*. Hong Kong: Government Press.

Hong Kong Government (1968), *Hong Kong Report for the Year 1966*. Hong Kong: Government Press.

Hong Kong Government (1969), *Hong Kong 1968*. Hong Kong: Government Press.

Hong Kong Government (1970), *Hong Kong 1969*. Hong Kong: Government Press.

Hong Kong Government (1971), *Hong Kong Report for the Year 1970*. Hong Kong: Government Press.

Hong Kong Government (1972), *Hong Kong Report for the Year 1971*. Hong Kong: Government Press.

Hong Kong Government (1973), *Hong Kong 1973, Report for the Year 1972*. Hong Kong: Government Press.

Hong Kong Government (1974), *Hong Kong 1974, Report for the Year 1973*. Hong Kong: Government Press.

Hong Kong Government (1975), *Hong Kong 1975, Report for the Year 1974*. Hong Kong: Government Press.

Hong Kong Government (1976), *Report for the Year 1975*. Hong Kong: Government Press.

Hong Kong Government (1977), *Hong Kong 1977*. Hong Kong: Government Press.

Hong Kong Government (1978), *Hong Kong 1978, A Review of 1977*. Hong Kong: Government Press.

Hong Kong Government (1979), *Hong Kong 1979, A Review of 1978*. Hong Kong: Government Press.

Hong Kong Government (1980), *Hong Kong 1980, A Review of 1979*. Hong Kong: Government Press.

Hong Kong Government (1981), *Hong Kong 1981: A Review of 1980*. Hong Kong: Government Press.

Hong Kong Government (1982), *Hong Kong 1982: A Review of 1981*. Hong Kong: Government Press.
Hong Kong Government (1983), *Hong Kong 1983, A Review of 1982*. Hong Kong: Government Press.
Hong Kong Government (1984), *Hong Kong 1984, A Review of 1983*. Hong Kong: Government Press.
Hong Kong Government (1985), *Hong Kong 1985, A Review of 1984*. Hong Kong: Government Press.
Hong Kong Government (1986), *Hong Kong 1986, A Review of 1985*. Hong Kong: Government Press.
Hong Kong Government (1987), *Hong Kong 1987*. Hong Kong: Government Press.
Hong Kong Government (1988), *Hong Kong 1988, A Review of 1987*. Hong Kong: Government Press.
Hong Kong Government (1989), *Hong Kong 1989, A Review of 1988*. Hong Kong: Government Press.
Hong Kong Government (1990), *Hong Kong 1990, A Review of 1989*. Hong Kong: Government Press.
Hong Kong Government (1991), *Hong Kong 1991, A Review of 1990*. Hong Kong: Government Press.
Hong Kong Government (1992), *Hong Kong 1992, A Review of 1991*. Hong Kong: Government Press.
Hong Kong Government (1993), *Hong Kong 1993, A Review of 1992*. Hong Kong: Government Press.
Hong Kong Government (1996), *Hong Kong 1996, A Review of 1995 and a Pictorial Review of the Past Fifty Years*. Hong Kong: Government Press.
Hong Kong Polytechnic (1989), *A Profile of New Students 1988*. Hong Kong: Student Affairs Unit, Hong Kong Polytechnic.
Hsieh, Jiann (1985), An old bottle with a new brew: The Waichow Hakkas' Associations in Hong Kong, *Human Organization*, 44 (2), 154–61.
Kao, Rosanna Santora (1987), Status of women in Hong Kong, *International Journal of Sociology of the Family*, 17 (1), 25–40.
Koo, Hagen and Hsiao, Hsin Huang Michael (1995), A comparative study of the East Asian middle classes: preliminary findings from South Korea, Taiwan, Hong Kong, and Singapore. American Sociological Association paper.
KPMG Peat Marwick (1989), *The Restructuring of Hong Kong's Manufacturing Sector: A Critical Transformation*. Hong Kong: KPMG Peat Marwick.
Kwan, Alex Yui-huen and Davis, Leonard F. (1994), Meeting the needs of the ageing population in Hong Kong: burden or challenge? *Hong Kong Journal of Social Work*, 28 (1), 66–76.
Kwok, Paulina, C. Y. (1995), Empowerment networking groups for women whose spouses are having extramarital affairs, *Hong Kong Journal of Social Work*, 29 (1), 60–64.
Lai, Lawrence Wai Chung (1993), Density policy towards public housing: a Hong Kong theoretical and empirical review, *Habitat International*, 17 (1), 45–67.
Lai, On-kwok (1995), The logics of slum formation in a capitalist-colonial regime – urban redevelopment hegemony reconsidered, *Guru-Nanak-Journal-of-Sociology*, 16 (2), 1–18.
Lau, Siu Kai (1981), Chinese familism in an urban-industrial setting: the case of Hong Kong, *Journal of Marriage and the Family*, 43 (4), 977–92.

Leete, Richard (1994), The continuing flight from marriage and parenthood among the overseas Chinese in east and southeast Asia: dimensions and implications, *Population and Development Review*, 20 (4), 811–29.

Lui, Ping-keung (1996), Public rental housing, familial processes and poverty: where is the Hong Kong underclass? Housing for millions – The challenge ahead. Conference Papers, Housing Conference 20–23 May 1996 (Ed. Hong Kong Housing Authority), 240–5. Hong Kong: Hong Kong Housing Authority.

Lui, Ting Terry (1983), Undocumented migration in Hong Kong specific measures taken to reduce the flow of undocumented migrants, *International Migration*, 21 (2), 260–76.

Luis, B. P. K., Chan, Kwok-leung and Choy, Yee-hung (1993), Determinants of the transition to marriage among young Chinese women in Hong Kong, *Hong Kong Journal of Social Work*, 27 (1), 34–41.

Luis, P. K., Cheung Chan-fai and Ho Wing-chung (1997), The relationship between Chinese morality and HIV/AIDS prevention: the ideational dilemma of men's promiscuity and gender equality (in Chinese). A paper presented at the Second Meeting of China – Hong Kong Committee on AIDS, Beijing.

Ma, Hong (ed.) (1996), *Connecting Shenzhen and Hong Kong to Create Mutual Prosperity* (in Chinese). Tienjian: Nankai University Press.

Ma, Joyce Lai Chong (1991), Meeting the changing roles of women: a call for a new policy strategy in Hong Kong, *Social Development Issues*, 13 (3), 126–37.

MacPherson, Stewart (1992), Social policy and economic change in the Asia Pacific region, *Social Policy and Administration*, 26 (1), 55–61.

MacPherson, Stewart (1993), Social security in Hong Kong, *Social Policy and Administration*, 27 (1), 50–7.

Mitchell, Robert Edward (1973), Residential patterns and family networks (II), *International Journal of Sociology of the Family*, 3 (1), 23–41.

Ng, Chun-hung (1997), *The economy restructured: women and the economic transformation of the Hong Kong-Greater China region*. American Sociological Association paper.

Ngan, Raymond and Wong, William (1994), *The caring injustice in family care of Chinese elderly people*. International Sociological Association paper.

Ozeki, Erino (1995), At arm's length: the Filipino domestic helper—Chinese employer relationship in Hong Kong, *International Journal of Japanese Sociology*, 4, Sept, 37–55.

Phillips, David R. and Bartlett, Helen P. (1995), Ageing trends – Singapore, *Journal of Cross-Cultural Gerontology*, 10 (4), 349–56.

Podmore, David and Chaney, David (1972), Attitudes towards marriage and the family amongst young people in Hong Kong, and comparisons with the United States and Taiwan, *Journal of Comparative Family Studies*, 3 (2), 228–38.

Podmore, David and Chaney, David (1973), Parental influence on the occupational choice of young adults in Hong Kong, and comparisons with the United States, the Philippines and Japan, *International Journal of Comparative Sociology*, 14 (1–2), 104–13.

Podmore, David and Chaney, David (1974), Family norms in a rapidly industrializing society: Hong Kong, *Journal of Marriage and the Family*, 36 (2), 400–7.

Pong, Suet-ling (1991), The effect of women's labor on family income inequality: the case of Hong Kong, *Economic Development and Cultural Change*, 40 (1), 131–52.

Pong, Suet-ling and Post, David (1991), Trends in gender and family background effects on school attainment: the case of Hong Kong, *British Journal of Sociology*, 42 (2), 249–71.

Post, David (1994), Educational stratification, school expansion, and public policy in Hong Kong, *Sociology of Education*, 67 (2), 121–38.

Post, David (1996), The massification of education in Hong Kong: effects on the equality of opportunity, 1981–1991. *Sociological Perspectives*, 39 (1), 155–74.

Raschke, Vernon (1976), Premarital sexual permissiveness of college students in Hong Kong, *Journal of Comparative Family Studies*, 7 (1), 65–74.

Rosen, Sherry (1978), Sibling and in-law relationships in Hong Kong: the emergent role of Chinese wives, *Journal of Marriage and the Family*, 40 (3), 621–8.

Salaff, Janet W. (1976a), Working daughters in the Hong Kong Chinese family: female filial peity or transformation in the family power structure? *Journal of Social History*, 9 (4), 439–65.

Salaff, Janet W. (1976b), The status of unmarried Hong Kong women and the social factors contributing to their delayed marriage, *Population Studies*, 30 (3), 391–412.

Schiffer, Jonathan R. (1991), State policy and economic growth: a note on the Hong Kong model, *International Journal of Urban and Regional Research*, 15 (2), 180–96.

Shek, Daniel T. L. (1996), The value of children to Hong Kong Chinese parents, *Journal of Psychology*, 130 (5), 561–9.

Tu, Edward Jow Ching and Lee, Mei Lin (1994), Changes in marital life cycle in Taiwan: 1976 and 1989, *Journal of Population Studies*, 16 July, 17–28.

Vagg, Jon (1993), Sometimes a crime: illegal immigration and Hong Kong, *Crime and Delinquency*, 39 (3), 355–72.

Wat, Sui Ying and Hodge, R. W. (1972), Social and economic factors in Hong Kong's fertility decline, *Population Studies*, 26 (3), 455–64.

Wilson, Christopher (1985), *The Dictionary of Demography*. Oxford and New York: Blackwell.

Wong, Fai Ming (1975), Industrialization and family structure in Hong Kong, *Journal of Marriage and the Family*, 37 (4), 985–1000.

Wong, Fai Ming (1977), Effects of the employment of mothers on marital role and power differentiation in Hong Kong, *International Journal of Sociology of the Family*, 7 (2), 181–95.

Wong, Siu-lun (1985), The Chinese Family Firm: A Model. *British Journal of Sociology*, 36 (1), 58–72.

Young, Katherine P. H. and Kwan, Roger W. H. (1995), The men speak out over their extramarital activities in China, *Hong Kong Journal of Social Work*, 29 (1), 47–59.

Index

Note: Page numbers followed by *n* indicate information is to be found in a note; page numbers followed by *f* or *t* indicate information is to be found in a figure or a table. Page numbers in bold indicate major discussion of a subject.